innovations in glass

The Corning Museum of Glass
Corning, New York

The Corning Museum of Glass
One Corning Glass Center
Corning, New York 14830-2253
(607) 937-5371

ISBN: 0-87290-146-7

Library of Congress Catalog Card Number 99-070716

Printed and bound in the U.S.A.

The Corning Museum of Glass
Corning, New York
Executive Director: David Whitehouse
Creative Director: Robert K. Cassetti
Editorial Coordinator: Richard W. Price
Graphics Coordinator: Jacolyn Saunders
Pre-press/Color Quality Supervisor: Nicholas Williams
Script Coordinator: Patricia Thiel
Content Coordinator: Connie Coburn
Image Researcher: Kristine Gable

Ralph Appelbaum Associates
New York, New York
Design Principal: Ralph Appelbaum
Project Director: Jack Pascarosa
Art Director: Matteo Bologna
Writer: Madeline Chinnici
Editorial Coordinator: Cheryl Filsinger
Content Coordinator: Jessica Holbrook
Graphic Designer: Fearn Cutler de Vicq de Cumptich
Design Liaison: Rick Sobel, Nadia Coen
Editor: Sylvia Juran

Printing and Binding
Upstate Litho, Rochester, New York

Preface

This book is the result of a transformation at The Corning Museum of Glass. Since opening its doors in 1951, the Museum has built and exhibited one of the world's richest collections of glass. The collection contains objects (in many cases, outstanding objects) from every period of glassmaking worldwide, and the Museum's galleries provide the visitor with a comprehensive introduction to the art and history of glass. That such a collection should be formed in Corning, New York, is not surprising. Glassmaking was America's first industry, and for more than 100 years, Corning has been the home of glassmakers. Indeed, in the late 19th century, Corning produced more brilliant cut glass than any other place in the United States, and received the nickname "Crystal City." In 1903, Frederick Carder came to Corning and, with local glassmaker T. G. Hawkes, founded Steuben.

Corning has done far more than supply the world with fine glass for collectors and connoisseurs, however. The first electric light bulbs were blown in Corning, and optical fibers were developed here. It is entirely fitting, therefore, that The Corning Museum of Glass should expand its exhibits to include the science and technology of glass, as well as its art and history. The Museum's Glass Innovation Center, opened in 1999, shows the visitor how, by exploiting the extraordinary properties of glass, innovators have changed the way we live. The light bulb made it possible for us to work and play in any of the 24 hours of the day. The television tube caused a revolution in entertainment and mass communications. Optical fiber has changed, and continues to change, the capacity of telecommunications. These advances, and the special properties of glass that made them possible, are documented in the Innovation Center.

This short book is not a guide to the Innovation Center. Instead, it focuses on the most important stories that are told in the Center: stories that revolve around discoverers of new kinds of glass, inventors of processes or products, the years of experimentation or the happy accidents that led to their discoveries and inventions, and how these innovations affect our lives.

The Corning Museum of Glass exists to tell the world about glass. This book will tell you why glass, which was discovered more than 4,000 years ago, is still one of the most remarkable and adaptable materials on earth—and in every vehicle launched into space.

David Whitehouse, Executive Director
The Corning Museum of Glass

Contents

Introduction

Glass is an ancient substance that has propelled us, leap by leap, into the future. We have curved it into lenses that revealed the richness of the heavens and the complexity of the microcosm, flattened it into panes that formed the glittering skin of skyscrapers, and pulled it into hair-thin fibers that sparked a revolution in global communications.

What is this powerful substance, this catalyst of civilization, that so easily bends to our will, and where does it come from?

Glass is not a single material. Almost any material—even a metal—can be transformed into a glass. A hybrid between a solid and a liquid, glass is a state of matter—the glassy state. Glass forms whenever materials (most commonly sand, lime, and soda) are melted and then rapidly cooled. During the heating stage, the molecules in the materials arrange themselves randomly, as in a liquid. As the melt quickly cools, it solidifies before the molecules have time to return to their ordered positions. Thus, glass has the random molecular arrangement of a liquid, but the rigidity of a solid.

The first glass was made by nature from lava that erupted from volcanoes, and from sand melted by strikes of lightning. And then, about 4,000 years ago, we discovered how to make glass ourselves. Not content with what nature had to offer, we changed glass, varying its chemistry and properties, and forming it in new ways—and it, in turn, changed us.

The new Glass Innovation Center at The Corning Museum of Glass offers a unique perspective on our long-standing relationship with glass. The exhibits—researched over a five-year period by staff in Corning and designed by Ralph Appelbaum Associates in New York City—convey life-changing stories of invention, creativity, and discovery. The exhibition, and hence this book, is organized into three galleries: Windows, Optics, and Vessels. Each gallery is an expression of one of the ingenious ways in which we have transformed and utilized the marvelous, versatile material we call glass.

Windows

Crown & Cylinder

making window glass by hand

When you think of a window, you think of a flat piece of glass. But window glass didn't always start out flat. It once began as a massive bubble on the end of a glassblower's pipe.

To flatten the bubble, the glassblower could spin it rapidly into a huge disk called a crown. After it was cool, the uneven, distorted disk was cut into small panes.

If, instead, the glassblower swung the bubble over a pit, he could stretch and elongate it into a cylinder. When the glass cylinder was cool, it was cut lengthwise, then reheated and flattened. Larger panes could be made this way, but the glass was still distorted—and the work was just as exhausting.

Glassmakers spinning a glass crown.

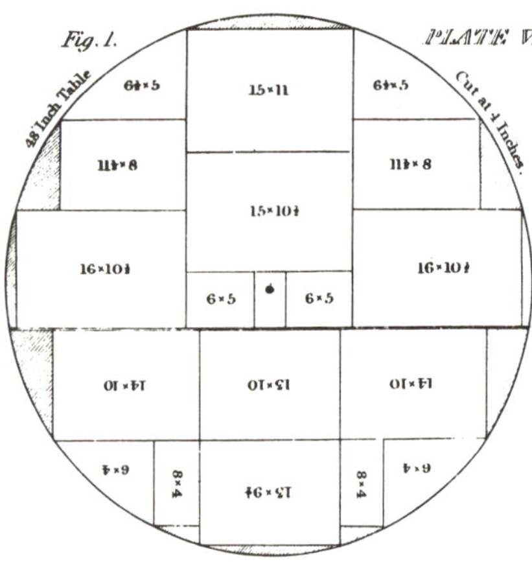

Glass crown marked for cutting.
Individual panes were cut after a glass crown cooled. The best panes were near the edge, where the glass was thinnest.

does window glass flow?

Some centuries-old window glass is thicker at the bottom than at the top. Is that because the glass flowed down over time?

No! Scientists have calculated that at room temperature it would take millions of years for glass to flow even a microscopic amount.

Older windowpanes have an uneven thickness because they were cut from uneven, handmade glass. The large glass disks made by the crown method, for example, were thicker in the middle than at the edge. Glaziers usually installed the thicker, heavier part of the pane at the bottom for stability.

19th-century cylinder glass workers.

pushing the envelope

All 245 designs had been rejected when horticulturist JOSEPH PAXTON submitted his radical, last-minute entry. It would become the most magnificent glass and iron structure ever built—the main exhibition hall of the first world's fair.

The task was enormous. The monumental building required 300,000 panes of glass, one-third of England's annual production. Each pane, made by hand using the cylinder process, was—

Glass

The history of architectural material ... has been the endless struggle for light ... in other words, the history of windows.

Le Corbusier, c. 1935

as Paxton had demanded—a foot longer than had ever been made before.

The Crystal Palace was erected in London's Hyde Park in 1851. Architecture has never been the same. The humble light-admitting glass pane had been transformed into a striking architectural material that captured the world's imagination.

how a glass cylinder becomes a windowpane

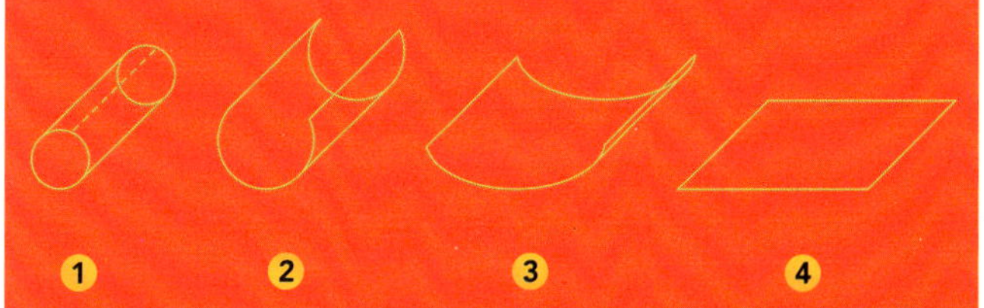

Interior view of the Crystal Palace, c. 1851.

the window machines
Sheet & Plate Glass

At the beginning of the 20th century, there was no way to mass-produce flat glass. Although glass cylinders could be drawn by machine, they had to be opened and flattened by hand. What was needed was a way to form sheets of glass directly and continuously.

EMILE FOURCAULT 1862–1919

In 1901, Belgian glassmaker EMILE FOURCAULT invented a machine that drew a glass sheet five stories straight up from a vat of molten glass. Now, sheet glass could be made more easily, but Fourcault's process marred its surface.

Sheet glass was good enough for most windows, but not for windshields. It distorted the view. Windshields required higher-quality plate glass, ground and polished to remove surface flaws. Making plate glass by hand was too slow to meet the growing demand for windshields.

By 1937, British glassmaker Pilkington Brothers had combined innovations with Ford Motor Company to develop a fully mechanized process for making plate glass. Twin-grinding was an engineering marvel, but the cost in energy and wasted glass was enormous.

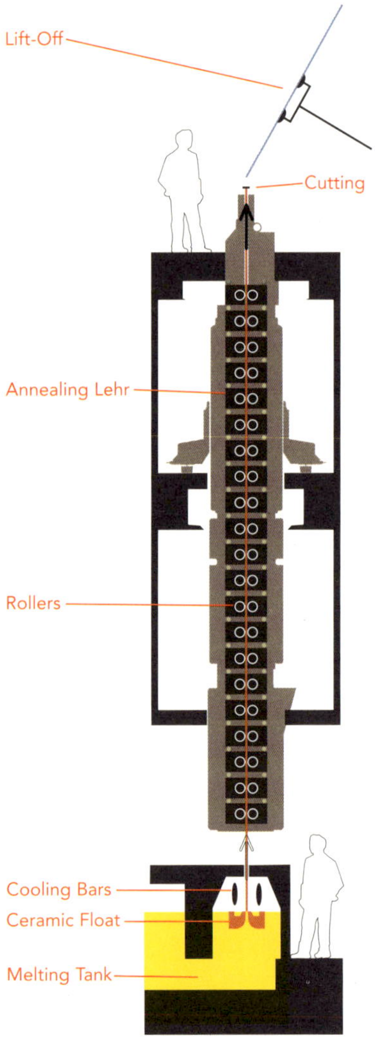

updraw window-making machine

Fourcault's updraw was the first machine to form flat glass directly. Glass was drawn through a slit in a ceramic float, then caught by rollers. The sheets were drawn five stories straight up from the vat.

Twin-grinding machinery.
The twin-grinding process was introduced in 1937. Rollers squeezed molten glass into a ribbon that was ground on both sides at once, then polished. The machinery stretched out over 1,400 feet (427 meters)—longer than the *Queen Mary*, then the largest ship afloat.

> The mighty glowing columns that stand like pillars in a ghostly cathedral ...
>
> Anonymous

Interior of a cylinder glass factory, c. 1910.
Giant glass cylinders were lowered with slings before being cut into windowpanes.

bigger cylinders

As long as glass cylinders were made by hand, their size was limited. Blowing a bigger glass bubble wasn't the answer. The glass would be too heavy to lift, too heavy to swing and elongate.

American glassworker JOHN LUBBERS found a way to make bigger window glass in 1902: he used a machine to draw a cylinder directly from a pot of molten glass. A circular bait, attached to a massive blowpipe, was lowered into the pot, then slowly raised. The glass followed, emerging as a spectacular cylinder that was ten times the size of a handmade one.

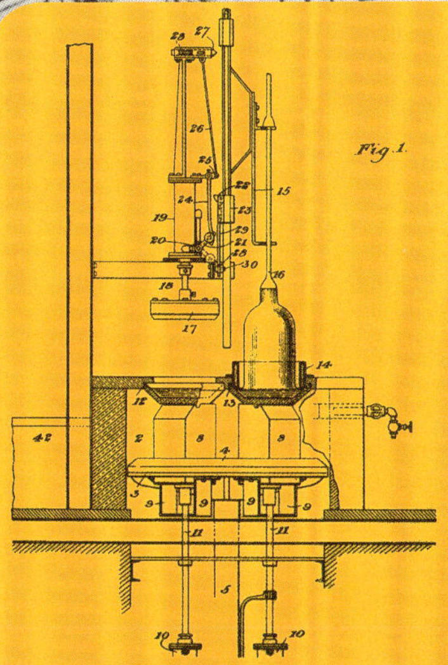

British patent 758,988
"Apparatus for Drawing Glass Articles" (detail)
J. H. Lubbers, May 3, 1904

Float Glass on a pool of tin

The auto and building industries boomed after World War II. Plate glass was in extremely short supply. Twin-grinding—used all over the world to make the glass—was too costly and too slow to keep up with the demand. Could high-quality glass be made without grinding and polishing?

SIR ALASTAIR PILKINGTON 1920–1995

ALASTAIR PILKINGTON, an engineer at Britain's Pilkington Brothers, was washing dishes when the idea struck: maybe it was possible to form a perfectly flat sheet by floating molten glass on a pool of liquid tin. That way, the glass would be smoothed by gravity and surface tension, rather than being squeezed by rollers. If the glass wasn't marred by rollers, it wouldn't need grinding and polishing.

The idea was brilliant in its simplicity, but seven costly years were needed to make it work. The risk paid off. The revolutionary Pilkington float process, introduced in 1959, is used today to make nearly all the flat glass in the world.

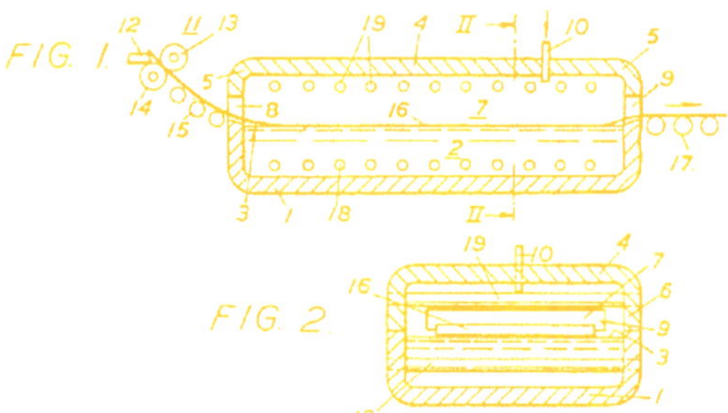

U.S. Patent 2,911,759
"Manufacture of Flat Glass" (detail)
L. A. B. Pilkington et al., November 10, 1959

making float glass

A float glass factory is tireless. Its tank can operate continuously for up to 12 years—making over 50,000 miles (80,000 kilometers) of high-quality flat glass almost automatically.

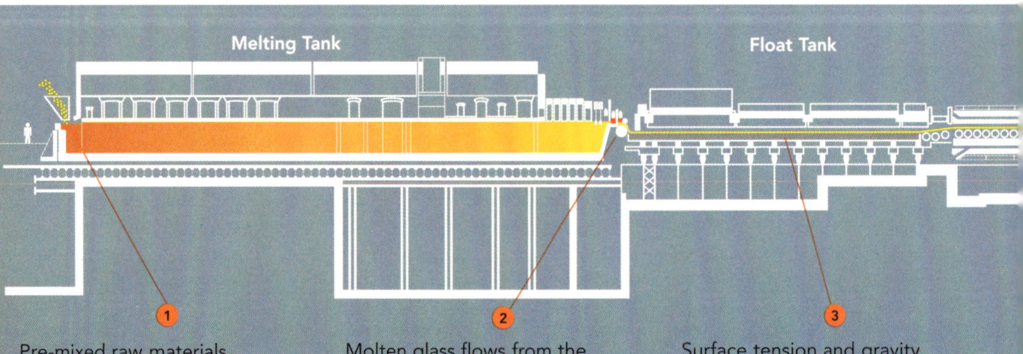

1. Pre-mixed raw materials continually move into a furnace, where they're melted.
2. Molten glass flows from the furnace over a spout onto the surface of a shallow pool of liquid tin.
3. Surface tension and gravity distribute the viscous molten glass into a relatively flat ribbon of uniform thickness, while edge rollers maintain the desired width.

> Had we been aware of the horrors which were to come, it is unlikely that we would have proceeded.
>
> Sir Alastair Pilkington, 1976

Float tank interior.
Floating on a pool of molten tin, hot glass is propelled by edge rollers from the inferno of the melting tank toward the annealer, cooling and hardening along the way.

A solid ribbon of glass emerges from the annealer.

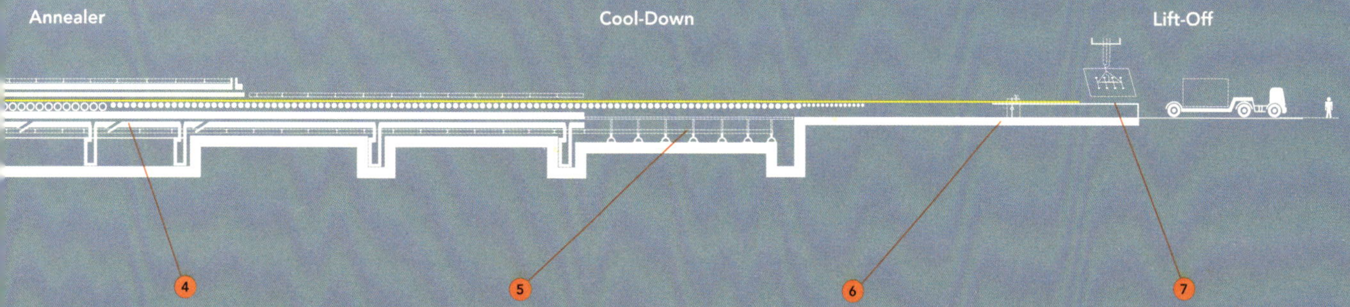

4 The solidified ribbon is cooled slowly in an annealing oven to prevent stresses from building up in the glass.

5 As the ribbon moves along, it further cools and hardens.

6 The edges of the ribbon, marred by the rollers, are trimmed off, and the glass is cut to size.

7 Robotic arms lift the glass panes and load them on a storage rack.

Tempered Glass

the precise moment

Glass breaks. But if it's strengthened by thermal tempering, it breaks less easily and more safely. By 1920, architects and European car designers wanted more and more tempered glass—and in large sheets. Glassmakers could successfully temper only one sheet in ten.

There's a trick to tempering: heat the glass until it almost softens and, at that precise moment, cool the surface rapidly and uniformly. Heat the glass too much and it becomes distorted, too little and it doesn't temper. Cool it unevenly and it breaks.

ACHILLE VERLAY, who worked at French glassmaker St.-Gobain, came up with a clever tempering solution in 1929. Using tongs with tips of tungsten steel, he hung a glass sheet in an oven. At the right temperature, the tips glowed, signaling that the glass was ready for cooling. Jets of air were then blasted uniformly on both sides of the glass, instantly tempering it. The process was nearly foolproof. It was automated and used for most of the 20th century to temper everything from car windows to basketball backboards.

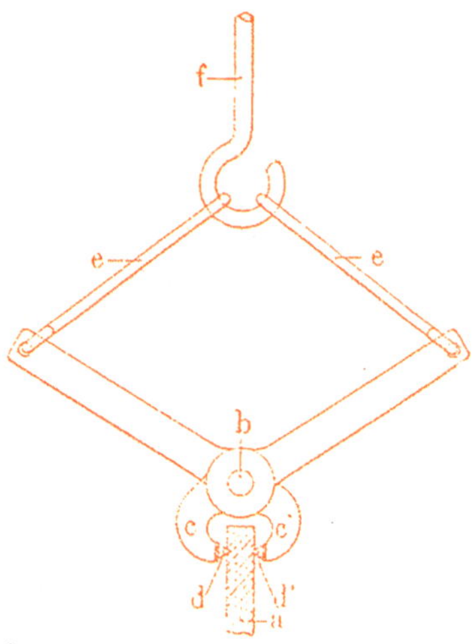

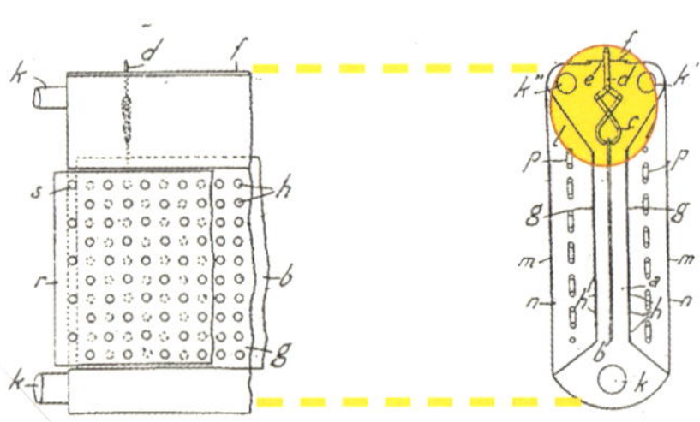

French Patent 38,676
"Procédé de trempe de glaces et verres" (details)
Société Anonyme des Manufactures des Glaces … de Saint-Gobain, October 11, 1932

the curious glass drops

PRINCE RUPERT OF BOHEMIA
1619–1682

PRINCE RUPERT OF BOHEMIA brought some curious glass drops from the Continent to the Royal Society in England in 1661. They had been made by dripping molten glass into cold water. The tadpole-shaped drops could be hammered repeatedly on their heads without breaking. Why were they so strong?

The drops remained a curiosity until, almost 200 years later, scientists discovered what was happening: when hot glass is quickly chilled, the fast-cooling outside layer is compressed. The slower-cooling inside is under tension. The only way to break the glass is to penetrate the tough outer layer and release the tension.

even stronger: chemically tempered glass

The windows of this space capsule have inner panes of chemically tempered glass for extra safety. Chem-tempered glass is 25 times as strong as ordinary glass. It's made by a chemical process that—like thermal strengthening—makes glass strong by compressing its surface.

The glass is placed in a bath of molten potassium salts. The combination of heat and salts causes the large potassium atoms to exchange places with smaller sodium atoms in the glass. As the surface of the glass becomes crowded, it becomes compressed.

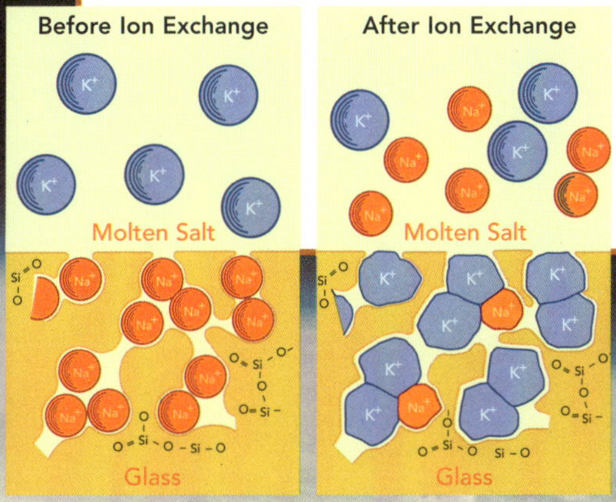

JESSE LITTLETON
1887–1966

In the 1920s, Corning physicist JESSE LITTLETON experimented to find the best temperatures for tempering glass and maximizing its strength. His discoveries gave glassmakers control over the tempering process and laid the groundwork for the flat-glass tempering method later developed by St.-Gobain.

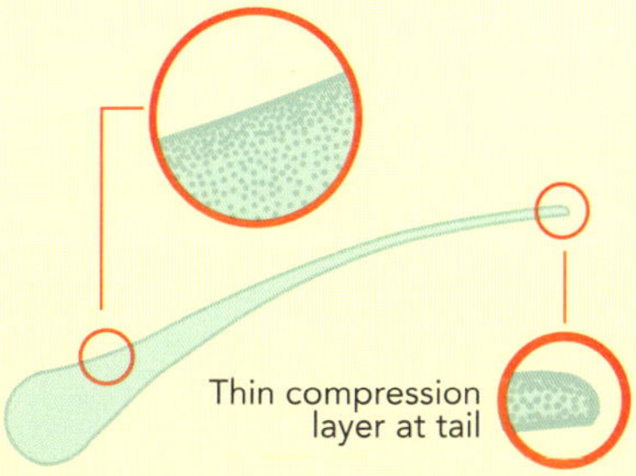

The head of a Prince Rupert drop is tough, but the tail is thin and can be broken. When the tail is snapped, all the tension trapped inside the drop is released, and the entire drop explodes into a fine glass powder.

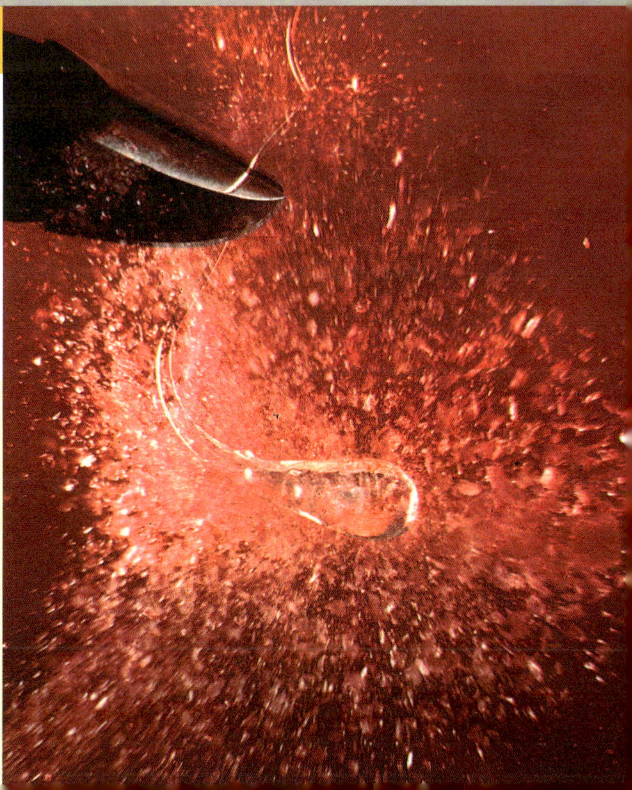

Laminated Safety from a broken flask

EDOUARD BENEDICTUS, a set and costume designer for a French theater, wanted to make glass safer. He was disturbed by reports of people being disfigured by broken windshield glass during automobile accidents. How could windshields be made less dangerous?

EDOUARD BENEDICTUS 1878–1930

He recalled a curious incident that had happened years earlier, when he was pursuing his interest in chemistry. He had dropped a seemingly empty glass flask, and it broke. To his surprise, the pieces stayed together. The liquid that had been in the flask had evaporated and left a thin plastic film, which held the pieces of glass in place.

Bénédictus concluded that windshields would be safer if they were made from glass that held together after it broke. In 1909, he patented TripleX™, a laminate made from two glass sheets bound to an inner layer of clear plastic. Today, every car, truck, and bus built in the United States has a windshield of laminated safety glass.

"The glass cracks, but it does not shatter." 1931 advertisement for windshields made with TripleX™. Illustration by A. M. Cassandre, a popular graphic artist of the period.

making a laminated windshield

A laminated glass windshield starts out as three distinct layers: two perfectly matched transparent glass sheets and an opaque plastic sheet.

The plastic interlayer is placed between the two pre-shaped glass sheets, and the entire assembly is subjected to heat and pressure. The plastic turns transparent and adheres to the glass, forming a three-layer windshield.

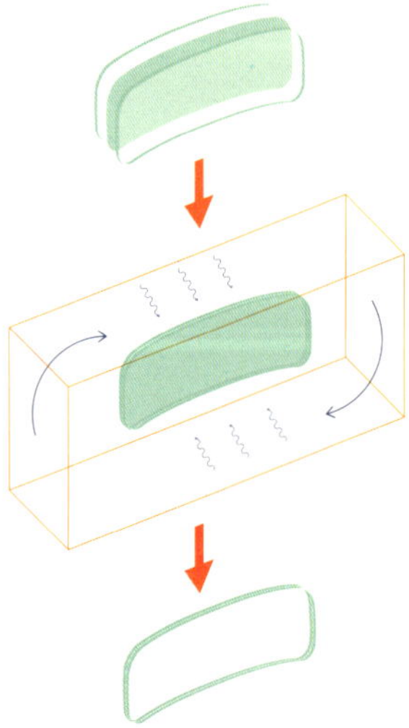

Glass

Early car with unlaminated windshield. Prior to the invention of laminated glass, an automobile accident could leave jagged glass shards in the windshield frame.

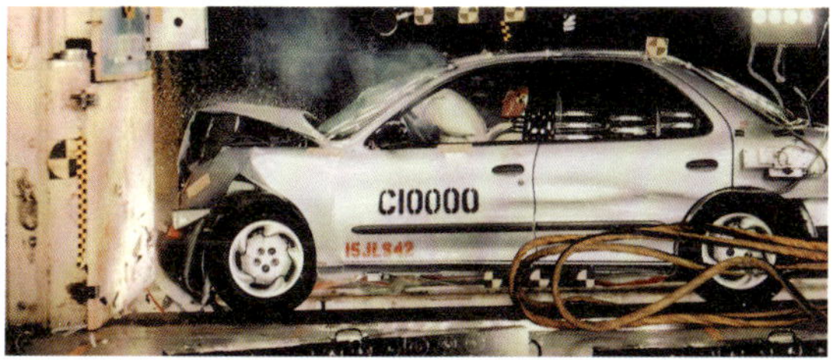

High-speed photographs of a crash test. Upon impact, the outside layer of the laminated windshield bends and breaks. The inside layer bulges and breaks with the pressure of the impact, but the plastic laminate keeps the pieces of glass in place.

Shaped Glass a pressing solution

Glass manufacturers had spent centuries learning how to make flat glass. Now, they wanted to bend it into complex shapes—without marring its surface. Anything that touched the surface of the hot glass could leave a mark.

The first curved windows were made by slumping. A glass sheet was placed in a gravity mold that touched only its edges, then was heated from above until it sagged into shape. Slumping worked for windows with simple curves, but it was of no use for making the more complex shapes that car designers demanded.

Then, General Motors asked for the seemingly impossible: an exotically contoured window for its new 1984 Firebird. There was no way to make the complex curves without touching the glass. American glassmaker Libbey-Owens-Ford had to take a bold step. It tried press-bending the precisely heated glass in an experimental full-contact mold made of a high-tech ceramic. The process worked. For the first time, glassmakers could totally control the shape of a window without marring its surface.

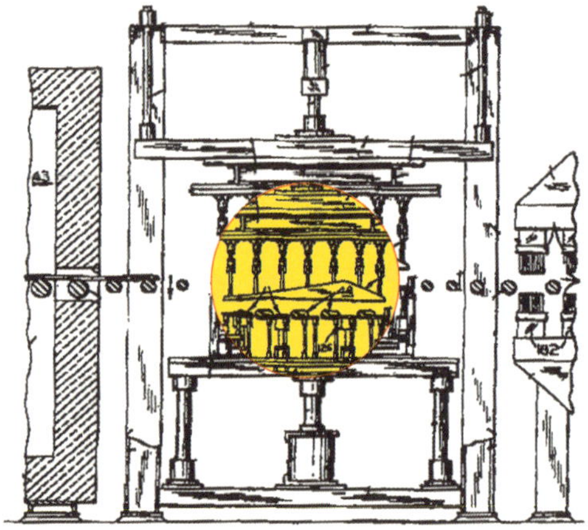

U.S. Patent 4,305,746
"Method of and Apparatus for Bending Sheets of Glass" (detail)
Libbey-Owens-Ford Company, December 15, 1981

1934 Chrysler Airflow, the first model with a curved, one-piece windshield. The first curved windshield presented an unexpected problem: two windshields broke for every one that was successfully installed. The rigid steel mounting that had worked well enough for flat glass was not sufficiently flexible for the new curved glass.

Press-bent Saab rear windows moving down the assembly line.

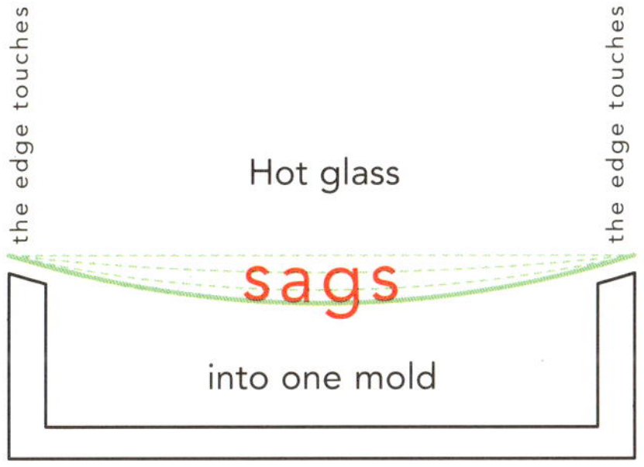

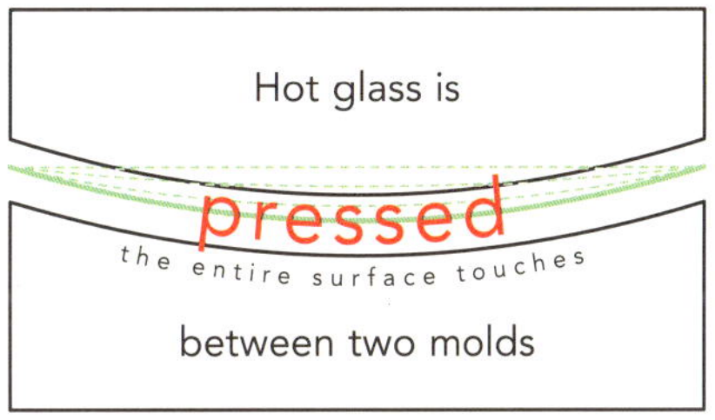

gravity window mold

In a gravity mold, the glass and mold are heated together. Each windshield travels through the furnace on its own mold, sagging into shape as it goes. To match the layers of a laminated windshield, two sheets of glass must be bent together, in the same mold.

press-bending window mold

Press-bent windshields are heated on their way to the mold. After being bent, each windshield leaves the mold, making way for the next. Today's press-bending process is so precise that windshields made separately have the perfect fit needed for lamination.

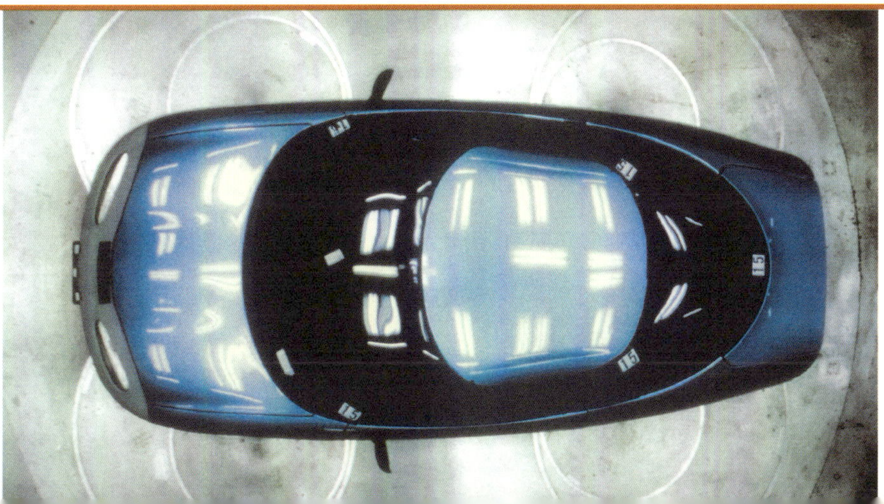

1999 GM EV1 electric car. This is the world's most aerodynamic production vehicle. The complex bends of the EV1's windows help reduce energy consumption by minimizing air "drag."

Fusion Draw Glass

the long road to success

Both companies were chasing the same thing: the enormous market for flat glass. While Pilkington was working to develop float glass, Corning Glass Works was developing a process of its own. Fusion draw produced a continuous sheet of exceptionally thin, pristine glass, but it was too slow to compete. Corning had a novel process without a product.

STUART DOCKERTY 1911–1974

CLINT SHAY b. 1922

The company used fusion draw first, in the 1960s, to make experimental windshields, then to produce optical-quality lenses for photochromic sunglasses. But the most demanding application didn't come until much later.

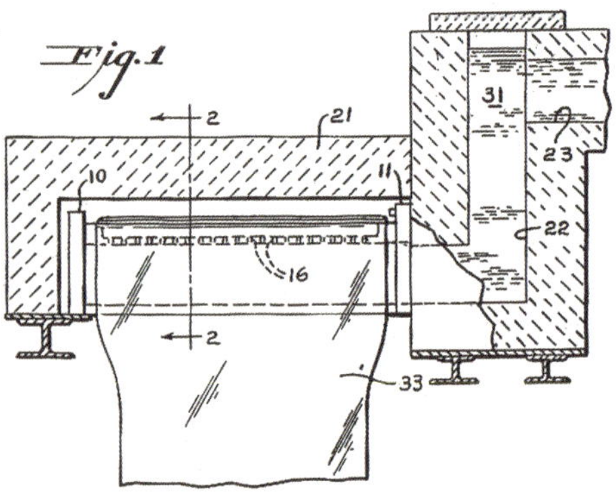

U.S. Patent 3,149,949
"Downflow Sheet Drawing Method and Apparatus" (detail)
S. M. Dockerty et al., September 22, 1964

In the 1980s, electronics manufacturers needed glass for liquid crystal displays (LCDs) in laptop computers. The glass had to be very thin and extremely flat, with a pristine surface on both sides. Only fusion draw could produce such flawless glass without grinding and polishing. Patience had finally paid off. Fusion draw became the process for making glass for the Computer Age.

how fusion draw works

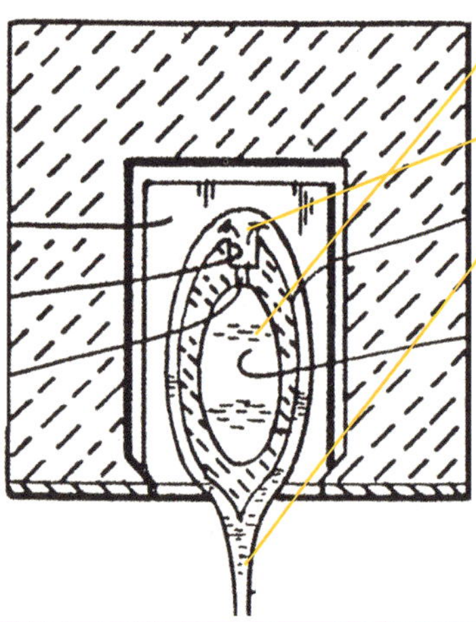

1. Molten glass feeds continuously into a tapered trough called an overflow pipe.
2. The trough overfills, and molten glass flows over both sides of the pipe in two streams.
3. The streams come together and fuse to form a single sheet. The glass that touched the surface of the trough is trapped on the inside of the sheet.
4. Edge rollers draw down the fused sheet. The edges of the sheet will be cut away. The remainder of the sheet has a pristine surface—nothing touched it while it was soft enough to be marred.

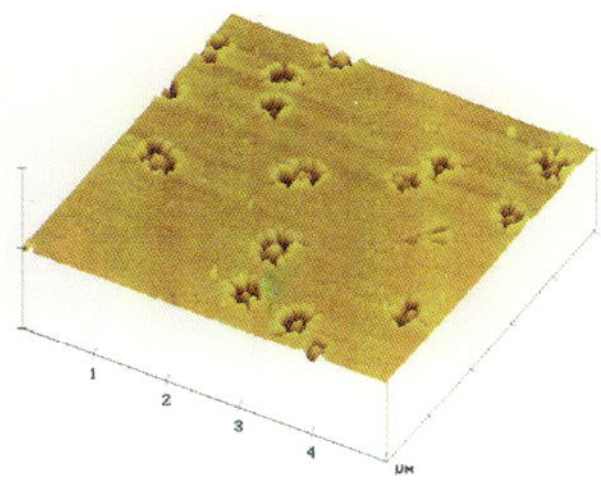

Float Glass Surface

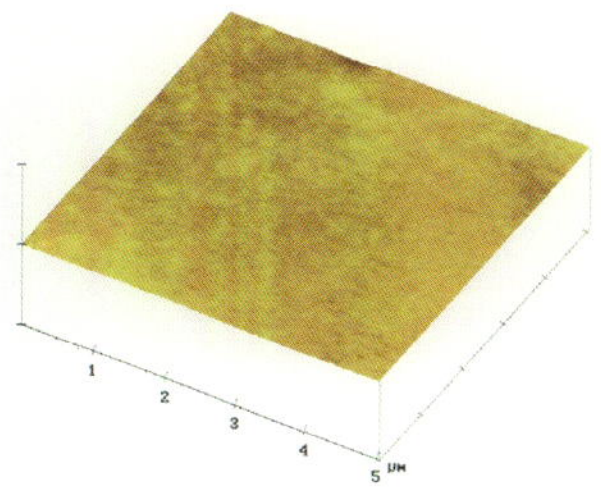

Fusion Draw Glass Surface

perfect on both sides

For some high-tech applications, such as active matrix liquid crystal displays (LCDs), glass must be perfect on both sides. Float glass is of very high quality—but it's not perfect. The side of the glass that touched the pool of tin is likely to have minor flaws. It's possible to polish float glass to perfection, but there's a better solution: fusion draw glass. It's untouched on both sides.

Fusion draw glass has been used to make optical-quality sunglass lenses, experimental car windshields, and LCDs for laptop computers. Without glass, laptop computer LCDs wouldn't be possible. Glass is the only economical, transparent substrate that can survive the high heat and harsh chemicals needed to lay down the display's electronics.

Coated Glass staying on-line

Glass windows are great, but they waste energy. A glass pane provides only slightly more insulation than nothing at all. Even so, builders were using more, and bigger, windows. Then, in 1973, the energy crisis hit.

ROY GORDON b. 1940

Harvard University chemist ROY GORDON knew that a tin oxide coating would make windows more energy-efficient. The challenge was in making the coating invisible. Gordon tried a two-phase chemical vapor deposition (CVD) process. He combined gases above a hot glass sheet. The gases reacted and deposited solid layers that formed a nonreflective coating of tin oxide on the surface of the glass.

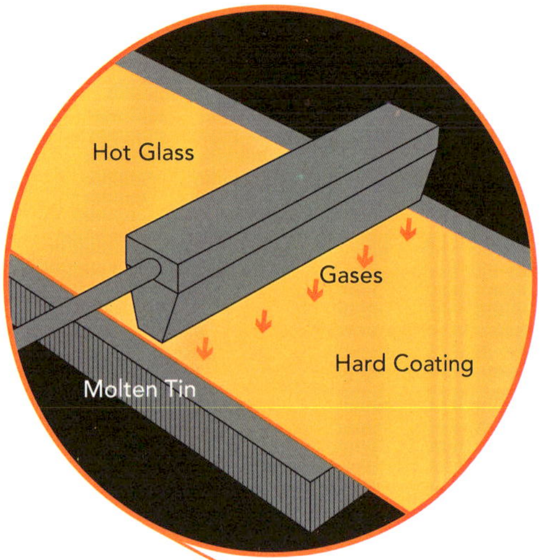

Gordon's innovation worked in the lab, but would it work in a float glass factory? The chemical reaction that formed the tin oxide fought the float process. Nevertheless, American glassmaker Libbey-Owens-Ford risked finding out. It took ten years, but in 1989, the company unveiled its on-line CVD coating process.

Energy-saving windows are not the only on-line coated glass we see around us. Coatings for reflective building glass, tarnish-free mirrors, and electrically conducting glass are all applied on-line, by CVD.

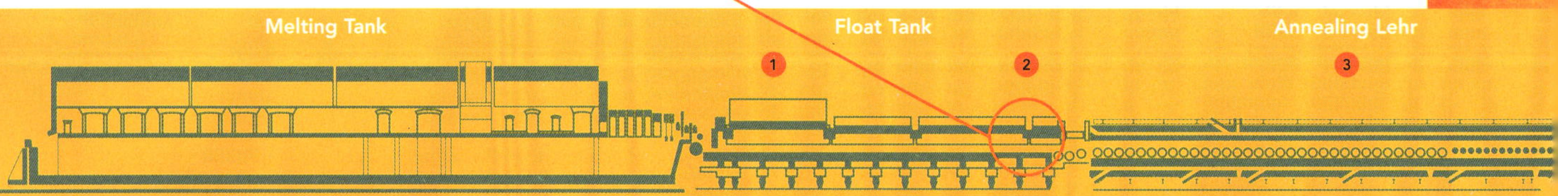

putting on a coat

There are several ways to coat glass, but on-line chemical vapor deposition (CVD) is the most efficient. Traditional processes apply a coating after the glass has left the float line. The continuous process coats the glass—while it's still hot—right on the float line.

1. Hot glass emerges from a furnace and spreads out into a continuous ribbon that floats on a bath of liquid metal. One or more coating machines, located above the ribbon near the end of the bath, release a mixture of gases.
2. Under high temperatures, the gases react to form a solid substance that deposits on the surface of the glass. The resulting coating is hard, uniform, extremely thin—and too tough to be scratched off easily.
3. The glass moves to an annealer, where it cools slowly.

blocking the heat

In the summer, a window coated with tin oxide blocks the sun's heat from going inside. The inside stays cool. In the winter, tin oxide prevents the heat inside from escaping. The inside stays warm.

Bank of China Tower, Hong Kong. Modern glass skyscrapers have windows with special coatings to increase energy efficiency.

Optics

Glass Lenses the quest to see more

A glass lens. It's nothing more than a curved piece of glass. So simple. So familiar. It's changed the way we perceive the world.

Leeuwenhoek's simple (single-lens) microscope. Leeuwenhoek's was not the first microscope, but it was better than any of the earlier ones: its tiny spherical lens gave it much greater magnifying power.

HANS LIPPERSHEY 1570–1619

ANTONJ VAN LEEUWENHOEK 1632–1723

In 1608, when Dutch spectacle maker HANS LIPPERSHEY held up two lenses, one behind the other, he was surprised to see a close-up view of a distant steeple. He had discovered a powerful tool. Soon, people all over Europe were putting lenses in tubes to make "spy glasses." Within a year, Italian scientist GALILEO GALILEI had made one, too. He used his instrument, which he called a telescope, to study the night sky. The moon, he saw, had rugged mountains; the Milky Way, countless stars.

Like the telescope, the first microscope had two lenses. Then, in 1674, Dutch naturalist ANTONJ VAN LEEUWENHOEK built the most powerful microscope of his time, using a single, spherical lens. He inspected a drop of water. It was teeming with life. His tiny lens had revealed that nature was far more complex than anyone had dreamed.

how Lippershey's telescope works

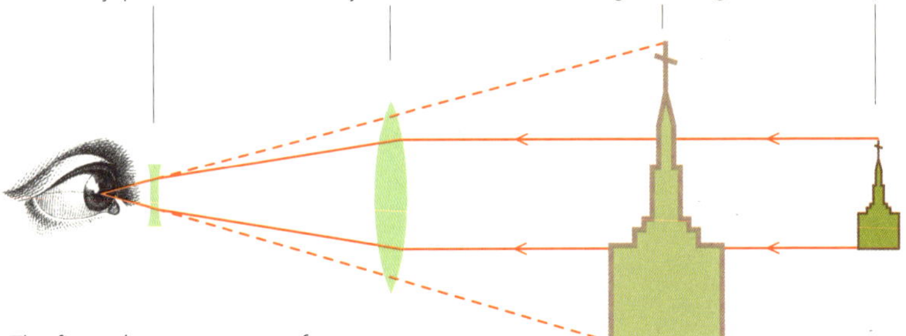

The first telescopes were refractors. They used lenses to bend and focus light.

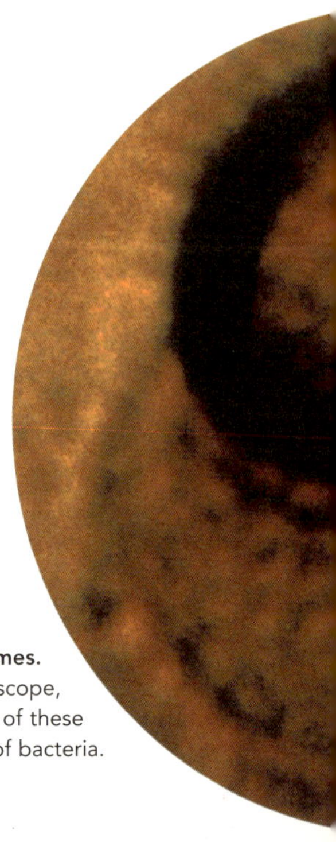

Truffle spores, magnified 600 times. This contemporary photograph, taken through Leeuwenhoek's microscope, shows how much Leeuwenhoek saw through his tiny lens. The size of these spores is that of many living cells. Their spines are the size of bacteria.

Where the telescope ends, the microscope begins.
Victor Hugo, *Les Misérables*, 1862

how does a lens bend light?

This lens is slowing down the light waves entering it from the air. The lens is denser than air—it slows light. As each wave front meets the curved surface of the lens, some parts slow down first, causing the light to bend, or refract. When the wave front leaves the lens, some parts speed up first, and the light bends again.

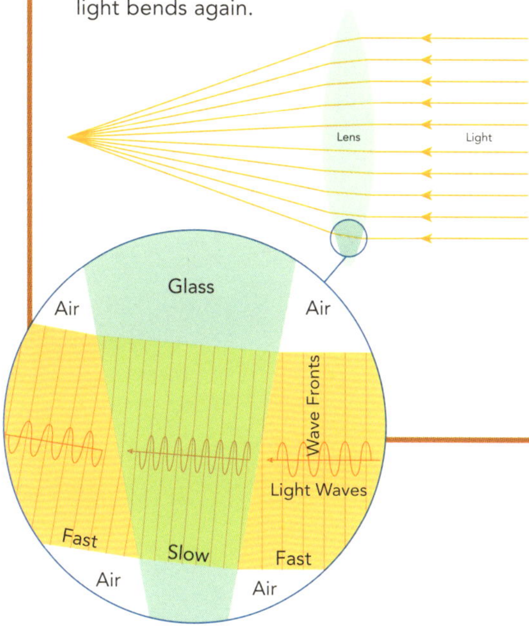

Galileo's telescope, c. 1630. Galileo's telescope gave him an unprecedented view of the heavens and a new insight into the universe. He became convinced that Earth and the planets revolved around the sun. Because of his ideas, Galileo was tried and convicted by the Inquisition in 1633.

Diagram by Galileo showing how to measure lunar mountains.

how a complex microscope works

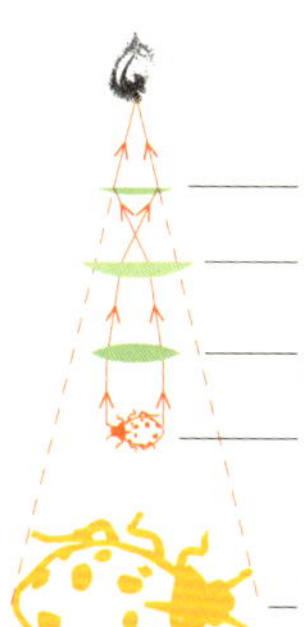

The first microscopes were natural outgrowths of telescopes, and had two lenses. British scientist ROBERT HOOKE's 17th-century compound microscope, like modern microscopes, has three lenses. An objective lens gathers light and forms an image. A field lens and an eyepiece lens then magnify the image.

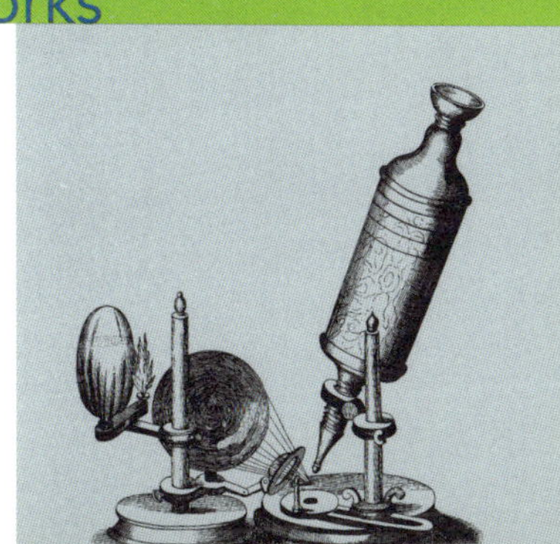

Hooke's microscope, 1665.

Telescope Mirrors

reflections on glass

The refracting telescope gave astronomers their first real glimpse of the heavens. Then, it began to frustrate them. At higher magnifications, the instrument's glass lenses produced distorted images. Pioneering British scientist ISAAC NEWTON solved the problem by using a metal mirror to gather light. He built his reflecting telescope in 1668.

Newton's reflecting telescope.

SIR ISAAC NEWTON 1642–1727

The more astronomers saw, the more they wanted to see. Reflecting telescopes became larger—until metal mirrors, too, turned out to be limited.

Astronomy might have reached a dead end if glass mirrors hadn't entered the picture. A glass mirror holds its focus better than a metal one because glass is less distorted by temperature changes. Glass doesn't corrode and it's easier to form into a precise shape. When French physicist Léon Foucault built the first large reflecting telescope with a glass mirror in 1864, he ushered in the age of modern astronomy.

pushing the limits

1939 cutaway rendering by Russell Porter of the Palomar Observatory.

GEORGE HALE 1868–1938

Some astronomers thought the universe was one grand galaxy. Others thought it contained many. In 1906, astronomer GEORGE HALE set out to build a telescope that could gather enough light to settle the debate.

Hale had already built the world's biggest reflecting telescope. Its 60-inch (1.5-meter) mirror was made from the largest piece of glass that had ever been cast. Still, it wasn't big enough. Making a larger one would be difficult, but that didn't stop Hale. After four painful years of pushing glassmakers to their limit, he had a mirror blank nearly double in size. When astronomer EDWIN HUBBLE used Hale's new 100-inch (2.5-meter) telescope to study the heavens, the debate was settled. Our galaxy is just one among many.

I contrived heretofore,
a perspective by Reflexion.
Sir Isaac Newton, c. 1668

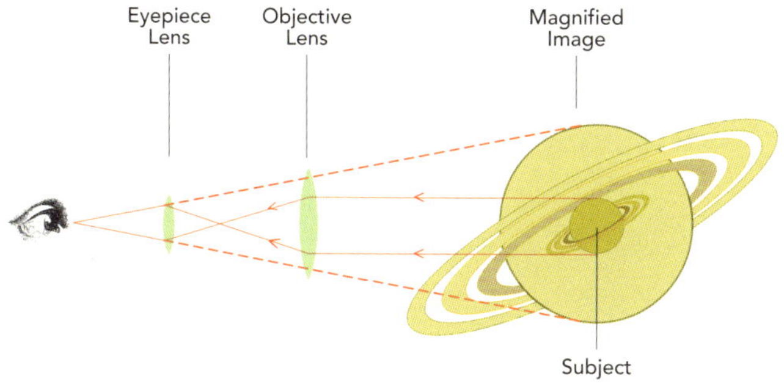

telescopes that refract light

A refracting telescope uses lenses to form an image. In astronomical refractors, both the large objective lens and the eyepiece lens are convex. The objective lens focuses light from a distant object, forming an inverted image. The eyepiece lens then magnifies the image. Astronomers see the images upside down, but the orientation doesn't matter to them.

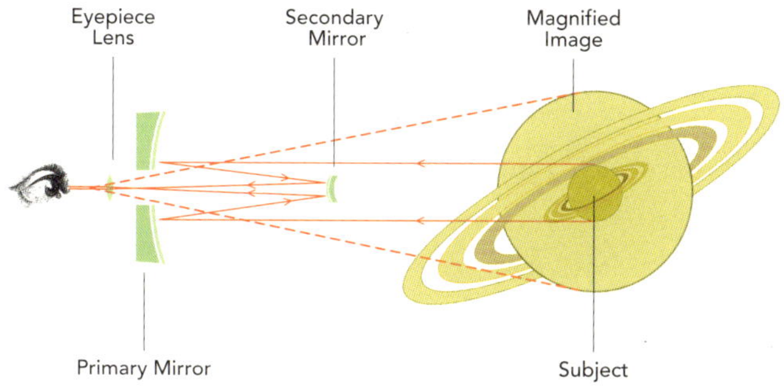

telescopes that reflect light

A reflecting telescope uses a curved mirror to form an image. The mirror acts as a light bucket to collect light from a distant object. In some reflectors, like the Hale, a smaller mirror captures the image and inverts it, sending it to a photographic plate or to an eyepiece lens that magnifies it.

200-inch Pyrex® mirror blank for the Palomar Observatory's Hale telescope.

Hale still wasn't satisfied. He wanted a 200-inch (5-meter) reflector. Would it be possible to make such an enormous mirror blank? In 1934, under the direction of physicist GEORGE McCAULEY, Corning Glass Works succeeded—on the second try. The new Palomar Observatory's giant Hale telescope stretched the boundaries of our knowledge, revealing an astounding universe made up of billions of galaxies.

Lens Combinations

countless variations

By the mid-1800s, there were still only two kinds of optical glass: soda-lime crown glass and lead-containing flint glass. Opticians doubted they'd ever have more choices. Then a German glass chemist made a crucial discovery. In 1884, OTTO SCHOTT proved he could precisely alter the optical properties of glass by changing its chemical makeup. The variations in glass, it now seemed, could be endless.

OTTO SCHOTT 1851–1935

ERNST ABBE 1840–1905

What sparked Schott's discovery was a plea for help published by a German physicist, ERNST ABBE. No matter how Abbe had configured microscope lenses, the image wasn't clear at high magnification. A colored fringe blurred the edge. The fringe, caused by chromatic aberration, occurs because a lens can't focus all colors of light to the same point. The only solution, Abbe had concluded, was to change the glass itself.

Schott sent Abbe samples of glasses he had made while experimenting with new recipes. After 93 tries, they found the right composition. Today, optical designers can correct for chromatic aberration and other distortions by carefully combining lenses made of different glasses.

Newton creating a rainbow with a prism.
In the 1660s, Isaac Newton observed that a glass prism spreads white light into a spectrum of colors. The effect, called dispersion, is beautiful—but it interferes with an image seen through the prism. In a lens, dispersion causes chromatic aberration, a fringe of color that blurs the image.

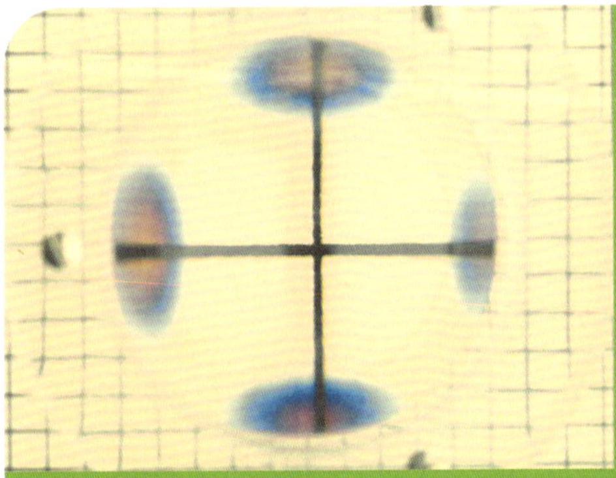

The multicolored fringe surrounding the edge of this image occurs because the lens can't focus all the colors of light to the same point.

> The world began to realize that so far it had only toyed with glass. Now a brand new material was born ...
>
> Walter Kiaulehn, *Odyssey of the 41 Glassmakers*, 1959

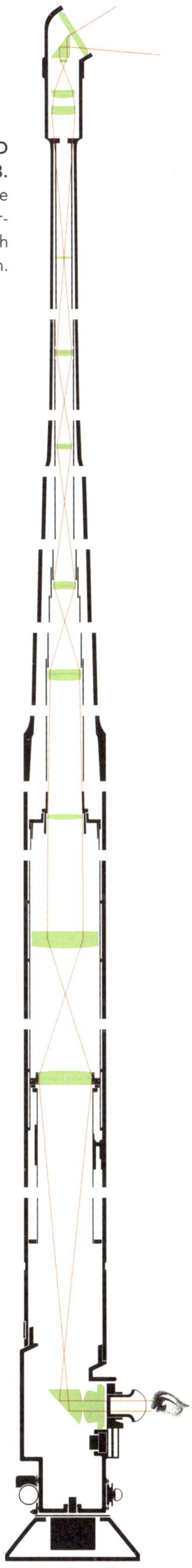

Section of Type 2D periscope lens assembly, c. 1943. High-performance optical devices have complex lens systems, with lenses of different shapes and compositions of glass. Each lens plays its part in minimizing aberration.

correcting the image

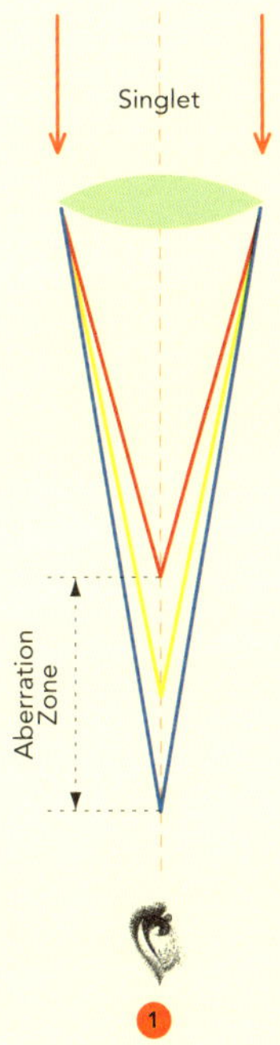

The convex lens focuses each color, or wavelength, of light at a different point. At high magnifications, the effect appears as a multicolored fringe around an image seen through a lens.

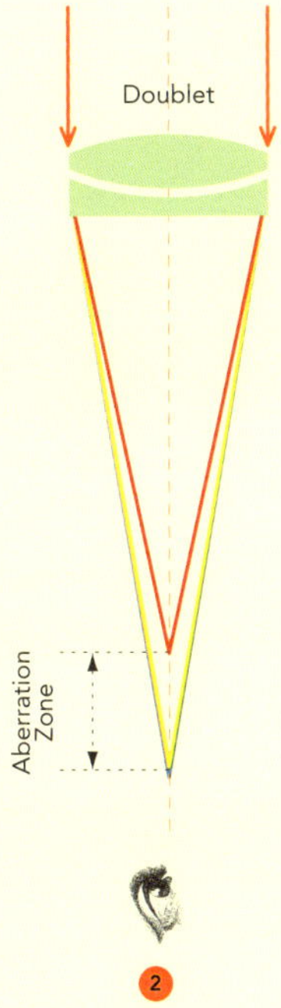

Adding a concave lens, made from glass with a greater ability to disperse light, focuses the red and blue light to the same point. The fringe partly disappears.

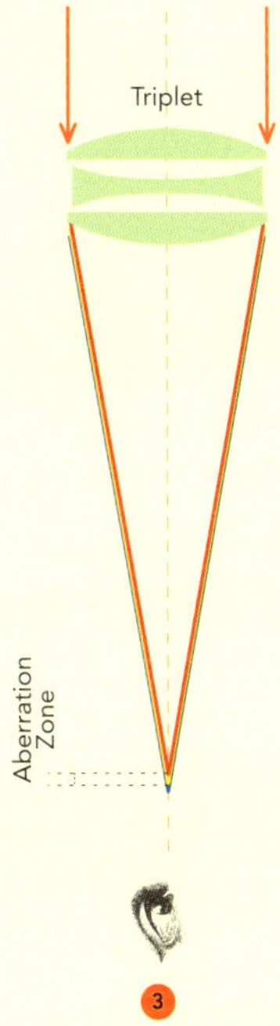

When another convex lens—made of yet a different glass—is added, the yellow light comes into the same focus as the red and blue light. Even more of the fringe is eliminated.

the troublesome rainbow

ISAAC NEWTON, the great 17th-century scientist, believed that the chromatic aberration inherent to a glass lens couldn't be corrected. With the single type of optical glass available to him, Newton didn't see a solution to the problem, and his authority was such that few continued to search for an answer. But in 1733, a partial solution was discovered—by an amateur astronomer.

CHESTER MOOR HALL had heard that lead-containing flint glass—discovered a few years after Newton had given up on lenses—had a greater ability to disperse light than ordinary crown glass. He ordered a concave lens made from flint glass, and a convex lens made from crown glass. When he stacked the two lenses, they partly canceled each other's chromatic aberration. Although traces of color remained, Hall's achromatic doublet was a dramatic improvement.

continuous perfection
Optical-Quality

To see the unseeable: the quest is unending. But lenses and prisms are only as good as their glass. Optical-quality glass must be flawless. Even tiny flecks, streaks, or bubbles can cause distortion.

CHARLES DeVOE
1909–1991

With World War II approaching, optical glass was in great demand, and making it was a slow, crude process. The ingredients were put in ceramic pots, melted, stirred, then cooled. Only about 10 percent of the glass was usable. Corning was among the glassmakers struggling to mass-produce optical glass in a continuous process.

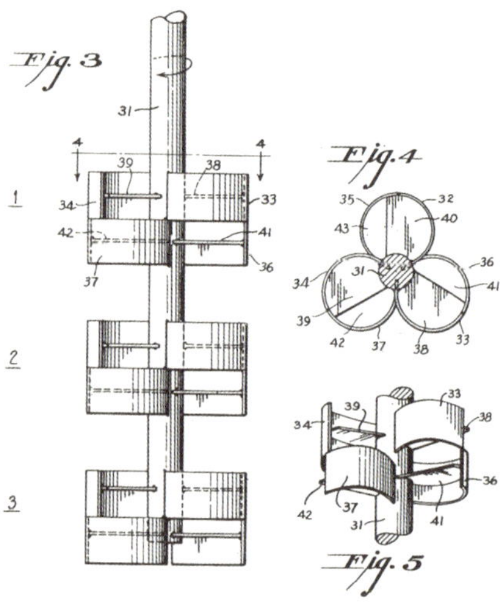

U.S. Patent 2,569,459
"Method and Apparatus for Stirring Glass" (detail)
C. F. DeVoe, October 2, 1951

Physicist CHARLES DeVOE knew that the problem holding back the mass production of defect-free glass was in the stirring. It was impossible to see what was going on inside a hot glass tank, so he built the world's first oil model of one. He used what he learned to design a delivery system with a platinum paddle-stirrer that thoroughly mixed the glass as it left the tank. By 1945, he had fitted his stirrer to an electric melting tank. Exceptionally high-quality optical glass flowed from the tank in a continuous stream.

not good enough

This piece of glass doesn't meet the criteria for optical quality because it contains defects:

Stones—tiny chunks or pieces of unmelted raw materials, small pieces of the melting tank that break off and contaminate the glass, or crystals that grow in the molten glass.

Seeds—bubbles that form during the melting process and become trapped inside the glass.

Striae or cord—streaks that aren't completely removed if the hot glass isn't thoroughly mixed.

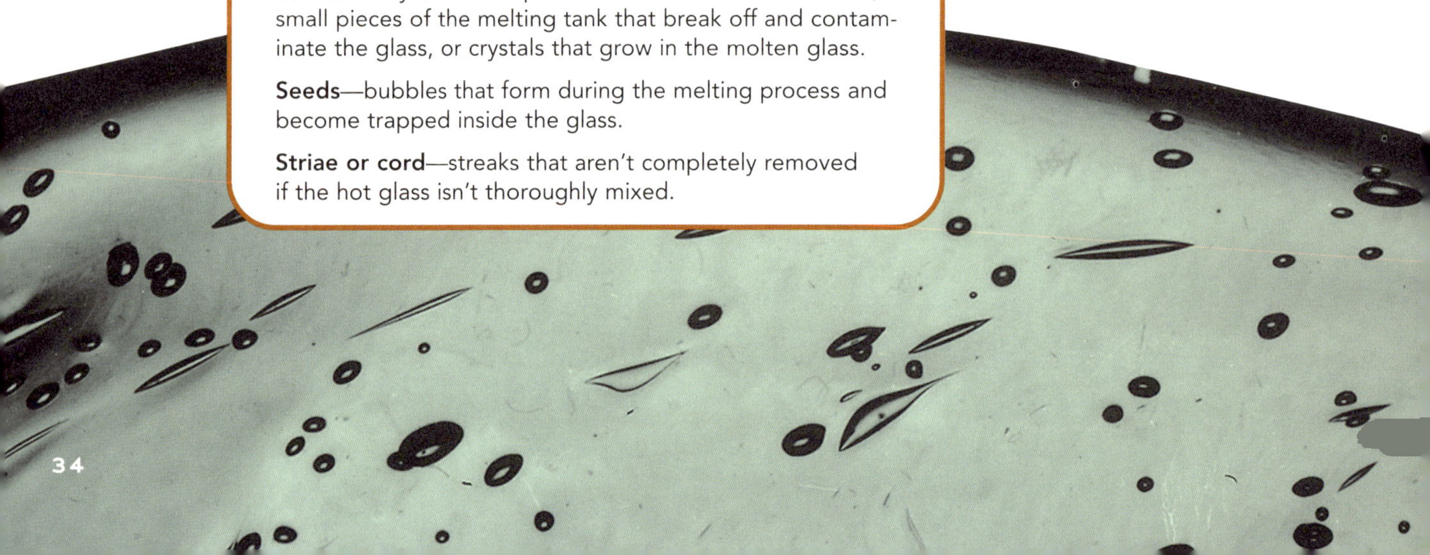

Glass

They make glass. By day and night, the fires burn on ... and bid the sand let in the light.

Carl Sandburg, *In Reckless Ecstasy*, 1904

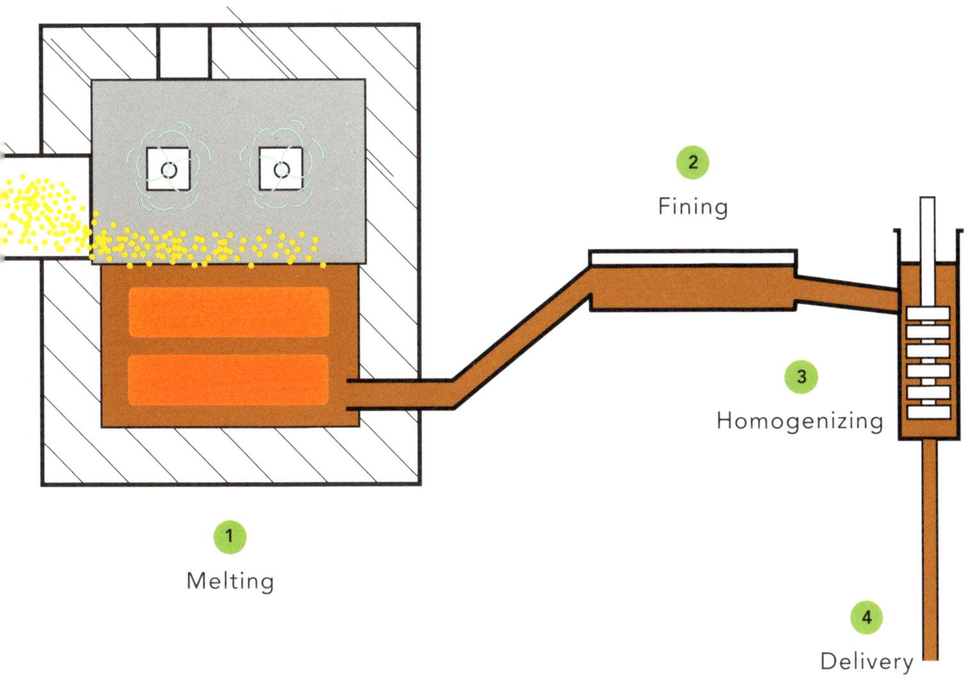

in a constant flow

1. Purified raw ingredients and recycled optical glass—called "batch"—are fed into the tank, which is lined with refractory ceramic. An electric furnace, sometimes aided by additional gas heat, turns them into molten glass.
2. The molten glass flows into a platinum fining chamber. There, it is heated even more, causing the bubbles in it to grow larger, rise to the surface, and break.
3. After passing into a delivery tube, the molten glass is homogenized—thoroughly blended—by a DeVoe-type platinum stirrer with curved paddles.
4. A continuous stream of optically pure glass emerges from the platinum delivery tube at a steady rate, night and day.

German glassworker breaking and sorting glass to find the clearest pieces for optical use.

Fused Silica: a break with tradition

It's exquisitely pure and remarkably transparent. It expands and contracts very little with changes in temperature. It is the simplest of all glasses, yet for years it was nearly impossible to make: pure silica glass. In 1934, Corning organic chemist J. FRANKLIN HYDE came up with a brilliant solution.

J. FRANKLIN HYDE
b. 1903

Glass had always been made by melting dry mineral ingredients. Making a pure silica glass required melting the purest quartz at incredibly high temperatures. Hyde's unprecedented idea was to synthesize the glass from liquid chemicals. In a crude experiment, he sprayed liquid silicon tetrachloride into the flame of a welder's torch. It reacted with the water vapor produced by the burning fuel to form an extremely pure glass—fused silica.

Hyde had invented a radical new way to make glass—the first in more than 4,000 years—and he hadn't even been given the assignment. Today, his process is used to make fused-silica glasses for the most demanding applications.

how fused silica is made

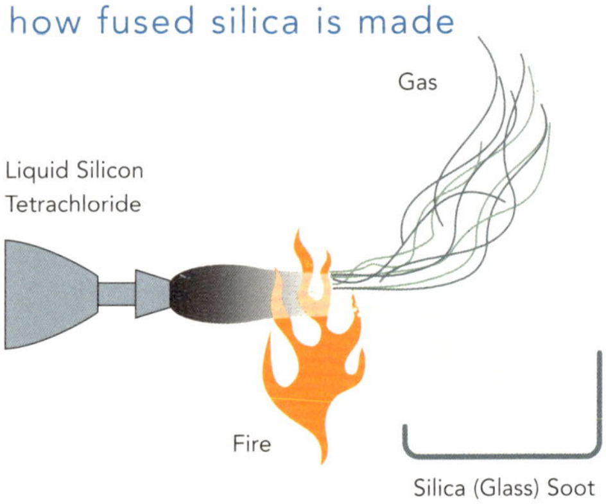

stepping down the size

A "stepper" lens assembly is the most sophisticated lens system in the world. It allows circuit patterns to be printed on computer chips like images on photographic film. Its dozen or more lenses, all made of fused silica, reduce a circuit's pattern and focus it—in lines as fine as a wavelength of light—onto the chip.

Fused silica transmits more ultraviolet (UV) light than any other kind of glass. The extremely short wavelength of UV light makes it possible to print nearly 9 million transistors on a silicon chip that is smaller than a postage stamp.

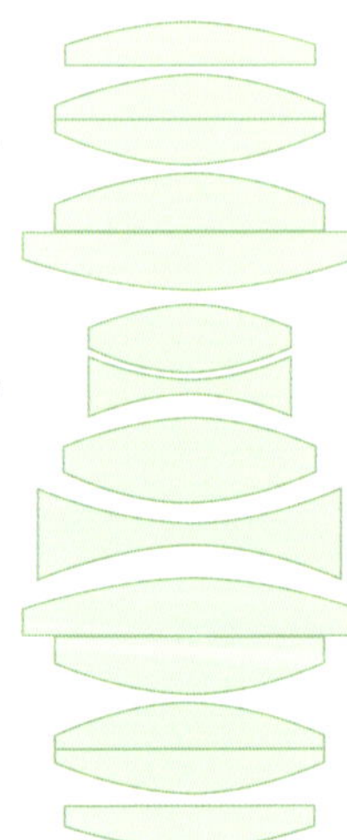

Space shuttle windows made from fused silica are tough and can withstand the tremendous heat of takeoff and re-entry.

After you see something work, then you realize that it's not so complicated after all.
J. Franklin Hyde

in cold and heat

Some telescope mirrors are made from Ultra Low Expansion (ULE™) glass—fused silica that contains the element titanium. Fused silica is very resistant to thermal changes, but titanium gives it greater thermal stability.

A telescope mirror made from ULE™ glass won't change its shape at all, even with extreme shifts in temperature. That's critical. Changes in a mirror's shape will distort the image.

Mirror for the Hubble space telescope, made from ULE™ glass.

Fresnel Lens
lighting the way

The most dangerous part of a sailing trip used to be returning to shore. Lighthouses were built to signal the safest route, but often the weak light from their lamps was not visible until too late. The large, thick lens that was supposed to project the light absorbed much of the signal. Hollowing out the interior of the lens helped, but not enough.

AUGUSTIN FRESNEL
1788–1827

French physicist AUGUSTIN FRESNEL knew that light rays spread out in all directions from their source. He spent eight years perfecting a system to capture those divergent rays and redirect them into a strong, concentrated, horizontal beam. In 1822, he unveiled his invention: a lighthouse lens made up of several thin lenses, each surrounded by concentric rings of prisms.

Travel has changed since Fresnel's day, but travelers still rely on Fresnel-type lenses for safety. They're used in everything from car headlights and traffic lights to airport beacons and ships' lights.

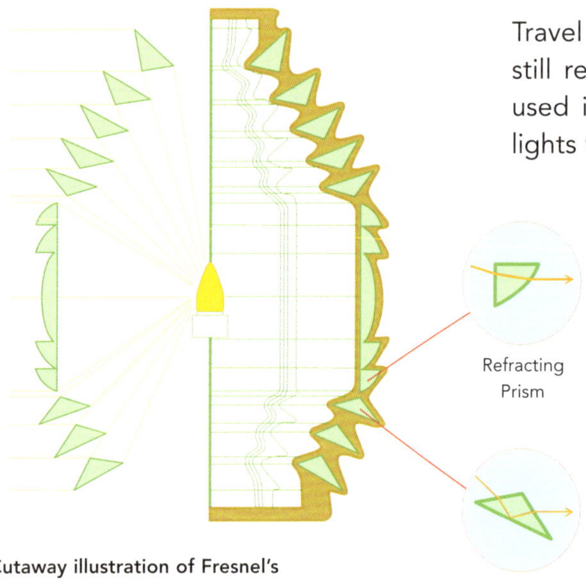

Cutaway illustration of Fresnel's lighthouse-lens system.
Some of the prisms in a Fresnel lens refract the light, allowing it to pass through in the desired direction. Some of the prisms reflect the light internally off one face, completely changing its course before it exits.

First-order Fresnel lens for the Makapuu Lighthouse, Hawaii, 1909.
The original Fresnel lighthouse lens was produced in seven sizes, or orders. The largest size was the first order, shown in the image here. A dozen or more people could stand inside it.

up to standard

By the 1890s, the railroad industry was booming. Fresnel-type signal lenses were being used to control the increasing train traffic. The trouble was that each rail line was using its own shades of each color for the lenses. Was a signal that appeared orange meant to be red—or yellow?

It was time to establish standards. Working with a color blindness expert, Corning chemist WILLIAM CHURCHILL identified the glass compositions that produced the shades of red, green, and yellow that the human eye finds most distinctive. The shades he developed would appear the same under any weather conditions. By 1908, the railroad industry had adopted Corning's colors as its standard.

Bundled Glass

Glass rods could transmit light, but could they transmit an image? A professor at a French agricultural college found himself faced with that question in the 1890s while he was tinkering with an early version of television.

HEINRICH LAMM 1908–1976

HENRI C. SAINT-RENE needed to find a way to transmit an image onto his display. A bundle of glass rods might work. Each glass rod would act as a pixel—a picture element—and together they would transmit an entire image.

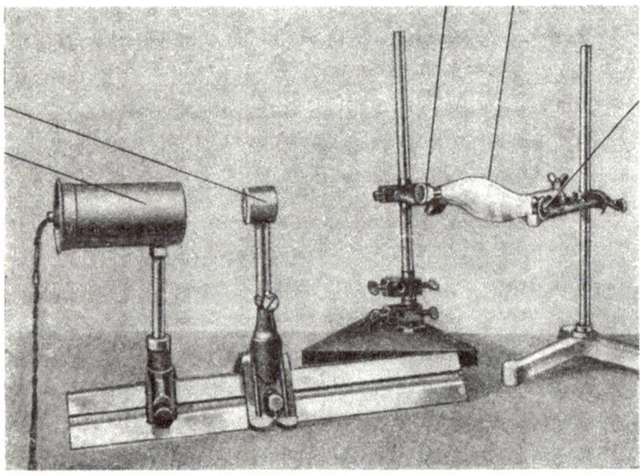

Lamm's illustration of his experiment, c. 1930.

Other inventors eventually came up with the same idea, but none tested it. Finally, in 1930, HEINRICH LAMM, a German medical student, proved that it worked. While trying to design a flexible gastroscope for peering inside the body, he put a light with a V-shaped filament at one end of a bundle of aligned thin glass fibers. When he looked at the other end, he saw a faint "V." Success! Lamm had transmitted the first fiber-optic image.

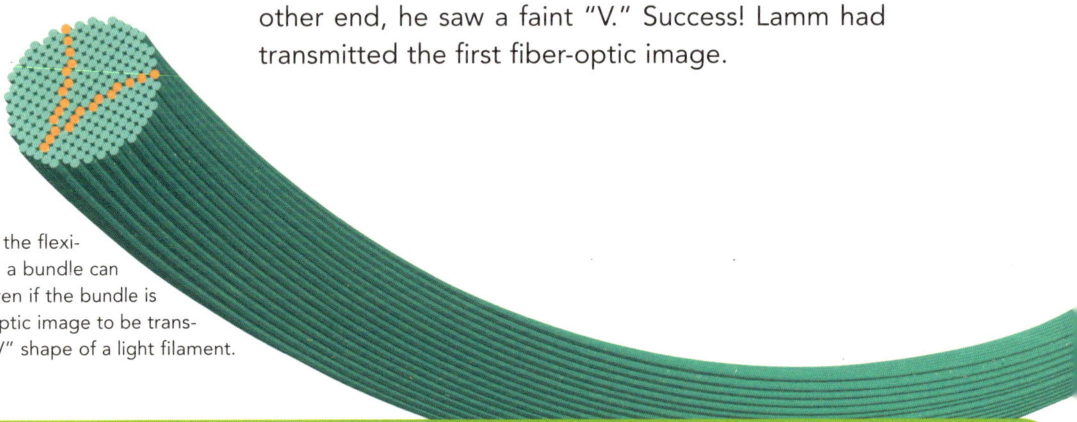

As long as they are aligned, the flexible, thin glass fibers inside a bundle can transmit an image—even if the bundle is twisted. The first fiber-optic image to be transmitted was the "V" shape of a light filament.

on the outside looking in

LAWRENCE CURTISS b. 1936

Imagine trying to swallow a rigid tube filled with lenses. It's not a pleasant thought—but doctors sometimes needed to peer down a patient's throat into the stomach. A flexible gastroscope, made from a bundle of glass fibers, would be easier to use, but the development of one that was practical had eluded scientists since the 1930s.

The more than 20,000 fibers had to be perfectly aligned; otherwise the image was scrambled. If the fibers became scratched or dirty, or if they touched one another, light escaped and the image was faint. Was coating the fibers the answer? The coatings that had been tried didn't work very well.

LAWRENCE CURTISS, a college student in Michigan, suggested coating the glass fibers with a different type of glass. His professor was sure the idea wouldn't work. But Curtiss proved him wrong. In 1957, he made the first flexible gastroscope, from fibers that were both aligned and coated with glass.

Fibers

around the bend

J. DANIEL COLLADON
1802–1893

JOHN TYNDALL
1820–1893

DANIEL COLLADON, a Swiss physicist, was an unlikely showman. But for years, European audiences were delighted with the illuminated water fountains that were based on his discovery. In 1841, Colladon had been experimenting with streams of water when he noticed that the light he used to highlight the streams followed the water's path, even around bends.

An Irish civil engineer, JOHN TYNDALL, described the phenomenon at the Royal Institution in London in 1854. The water traps most of the light, guiding it along by total internal reflection. The light can't escape because it's reflected whenever it hits the boundary between the water and the surrounding air at a shallow angle.

When the effect was demonstrated using a bent glass rod instead of a stream of water, the groundwork was laid for guiding light to and from otherwise inaccessible places.

total internal reflection

The smooth surface where the stream of water and the air meet acts like a mirror. When the light strikes it at a shallow angle, it reflects light better than any man-made mirror. Total internal reflection traps the light in the stream—and will trap it the same way in a glass fiber.

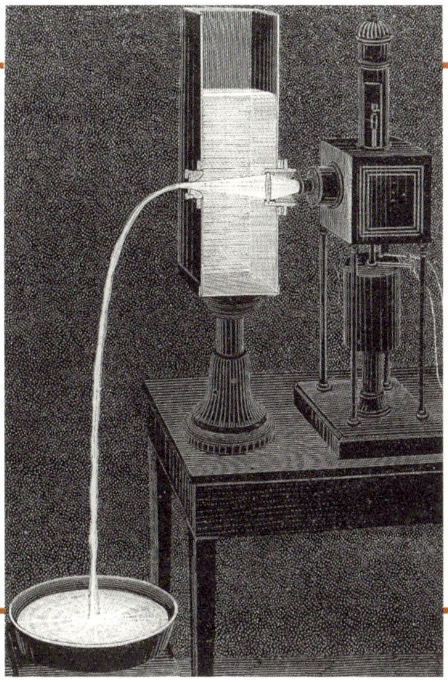

Illustration of Colladon's fountain. *Scientific American*, 1889.

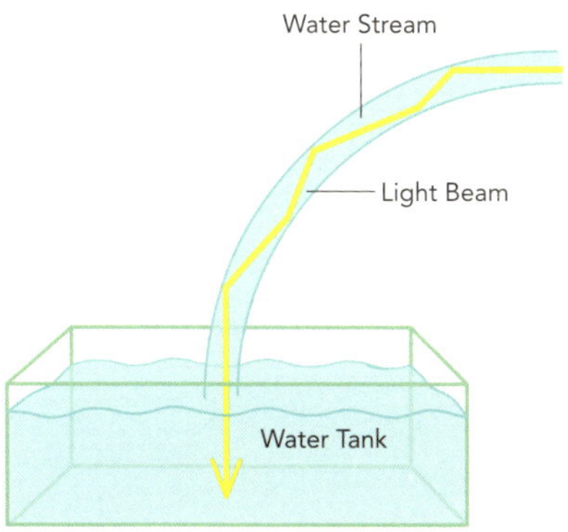

Optical Fibers for

on a thread of glass

For nearly a century, scientists had dreamed of using light to "talk" long distances. Light can carry a lot of information, but it needs something to guide it. Mirrors and lenses had been tried, but had proved impractical. There seemed only one choice: glass fibers.

DONALD KECK b. 1941

ROBERT MAURER b. 1924

PETER SCHULTZ b. 1942

The problem was that fibers made from ordinary glass absorbed too much of the light signal. In 1964, CHARLES KAO, an engineer at Standard Telecommunications Laboratories in England, made a critical calculation. If the impurities were removed from glass, enough of the signal would get through to make an optical communication system a reality. Around the world, scientists raced to purify glass and use it to make optical fiber.

Corning Glass Works already had the world's purest glass: fused silica. Physicist ROBERT MAURER decided to use it to form the cladding, or skin, of the fiber. With the help of chemist PETER SCHULTZ and physicist DONALD KECK, he made the fiber's core from fused silica with the added element titanium. The "extra" element changed the optical properties of the core glass in a way that ensured that the light would be totally reflected inside the fiber.

On an August day in 1970, Keck set out to measure the latest prototype. When he looked at the end of the fiber through a microscope, the light was so bright, he said, that "it hit me in the eye." That first low-loss optical fiber sparked the telecommunications revolution—and catapulted us into the Information Age.

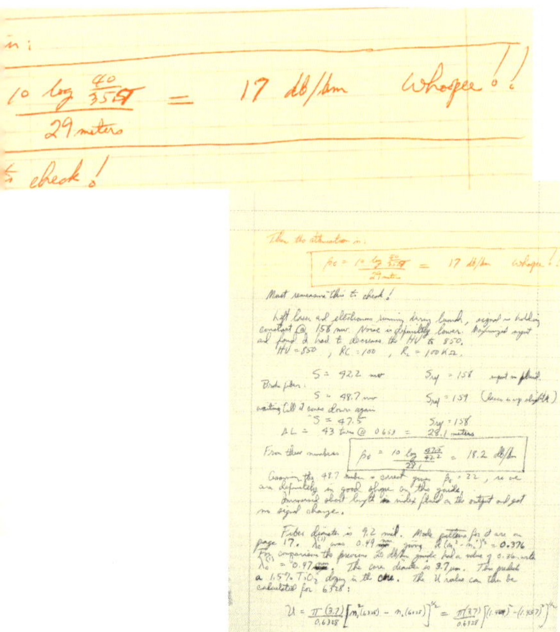

Original notebook entry for the first low-loss optical fiber. When Donald Keck discovered that Corning's latest optical fiber prototype had a low signal loss, he was so excited that he recorded his unscientific reaction in his laboratory notebook. Later, he retested the fiber and found the signal loss was even less than he had thought.

using light to talk

It was a bright idea: use sunlight to transmit the human voice. In 1880, American innovator ALEXANDER GRAHAM BELL tried it, using a thin, flexible mirror to reflect a light beam onto a distant receiver. His voice caused the mirror—and the reflected light—to vibrate. The receiver detected the light's vibrations and turned them back into sound.

Bell's photophone worked, but it wasn't practical. Rain, smoke—anything in the atmosphere—scattered the unpro-

Communication

I have heard a ray of the sun laugh and cough and sing!

Alexander Graham Bell

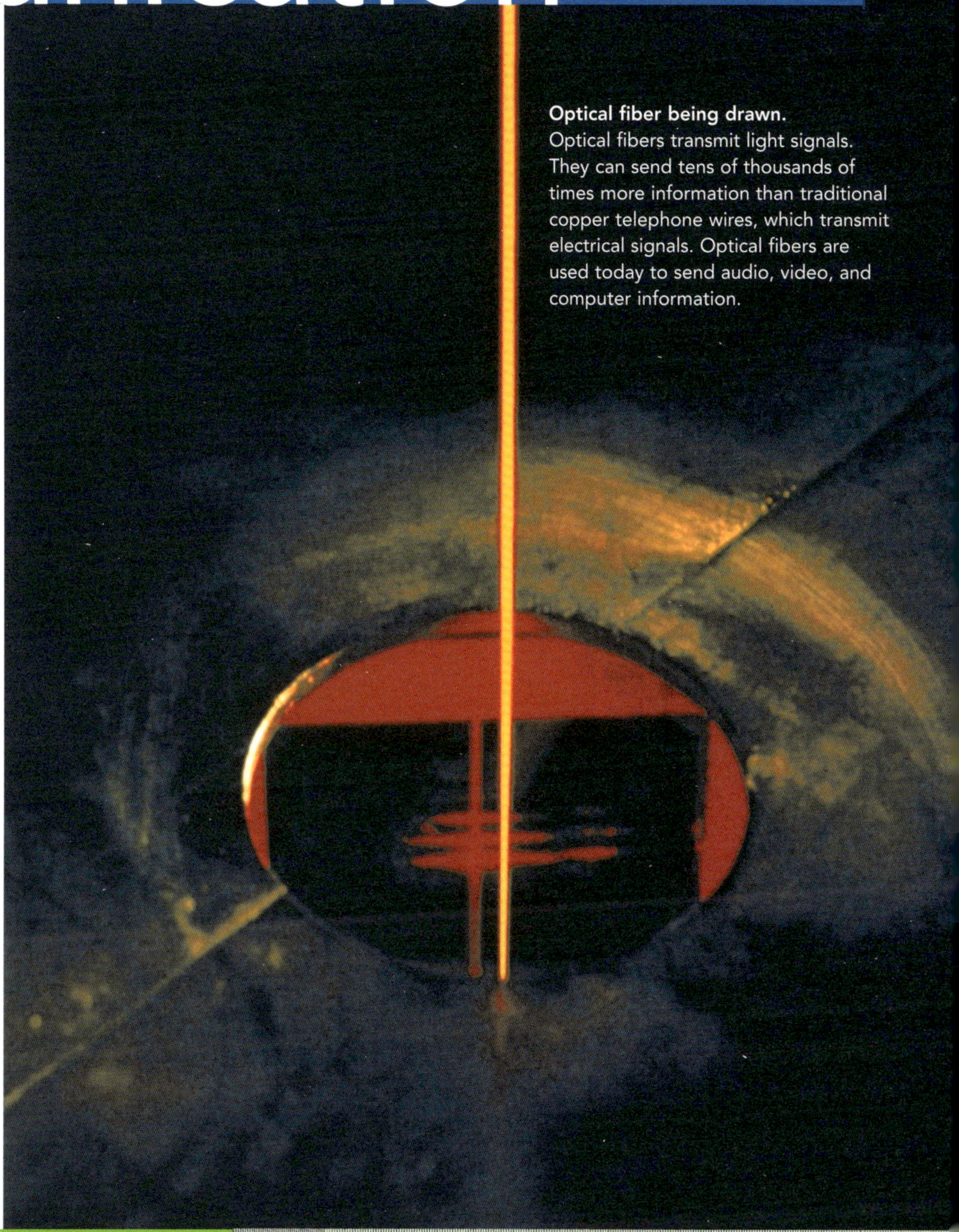

Optical fiber being drawn.
Optical fibers transmit light signals. They can send tens of thousands of times more information than traditional copper telephone wires, which transmit electrical signals. Optical fibers are used today to send audio, video, and computer information.

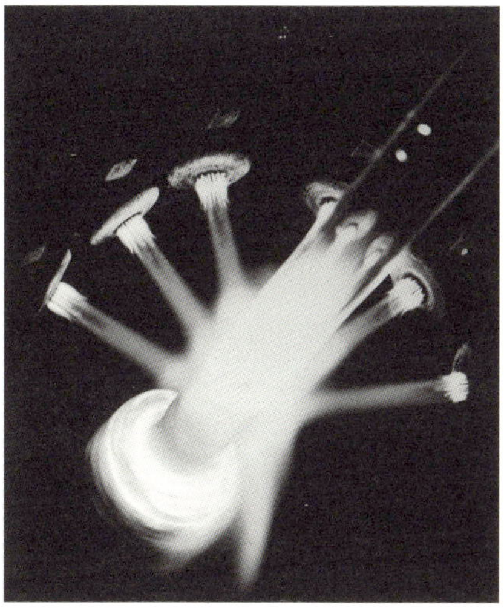

Most optical fibers are made using a process called outside vapor deposition. First, a rod is sprayed with the core glass, then cladding glass is sprayed over the core to form what is called a blank. The rod is removed, and the blank is heated to consolidate the glass. The fiber that's pulled from the blank has the same core-cladding geometry as the blank.

tected light signal. Bell's idea languished. Instead of sunlight, microwaves and radio waves, which are less affected by the atmosphere, became the media of choice.

As the demand for communication grew, the problem with microwaves and radio waves became apparent: they couldn't carry enough information. Shorter-wavelength light could carry more. Eventually, Bell's idea of optical communication resurfaced, as did the problem of how to protect the light.

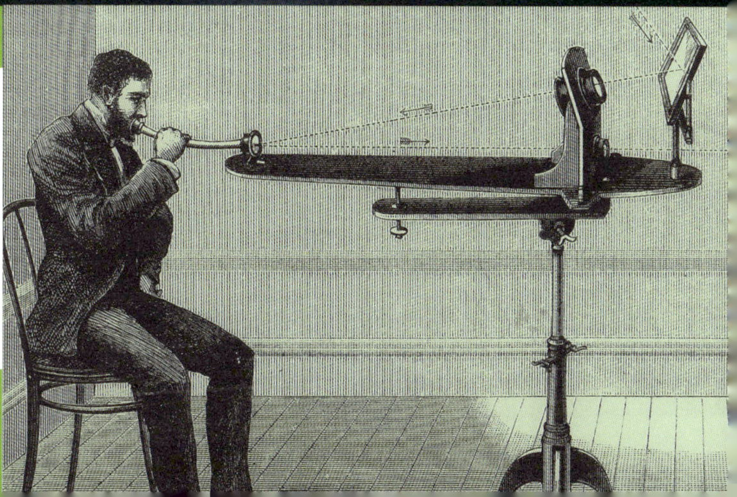

Transmitting end of Bell's photophone.

Glass That Amplifies

with a burst of energy

It was an idea that might have remained buried in scientific journals. Instead, it led to a device that gave modern telecommunications a much-needed boost.

ELIAS SNITZER
b. 1925

In the 1960s, ELIAS SNITZER, a physicist at American Optical, added rare earth elements to glass. These elements can absorb light energy—and, he discovered, also amplify it, emitting large amounts of power at a single wavelength. Snitzer had invented the glass laser. Other lasers were more practical for most uses, but Snitzer realized that his had one advantage: glass that amplifies light—an optical amplifier—could be made in any size and shape.

Twenty years later, telecommunications was at an impasse. The electronic devices used to boost light signals traveling through optical fibers had reached their limit. Glass optical amplifiers were just what was needed to meet the demand for more and better communication. Faster and more powerful, they could efficiently amplify light signals up to 10,000 times.

Glass optical amplifier. No matter how perfect an optical fiber is, a message-carrying light signal traveling through it eventually loses power. Glass optical amplifiers are used to strengthen the signal.

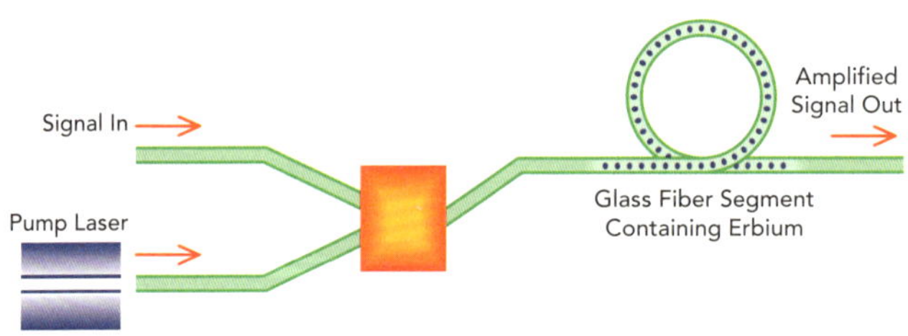

how an optical amplifier works

An optical amplifier consists of a tiny "pump" laser and a short segment of glass fiber that contains the rare earth element erbium. The erbium absorbs the pump laser's energy and stores it until the passing signal stimulates its release. The newly released light energy mimics the message-carrying signal. The extra light is then added to the signal as it moves on, strengthening it up to 10,000 times.

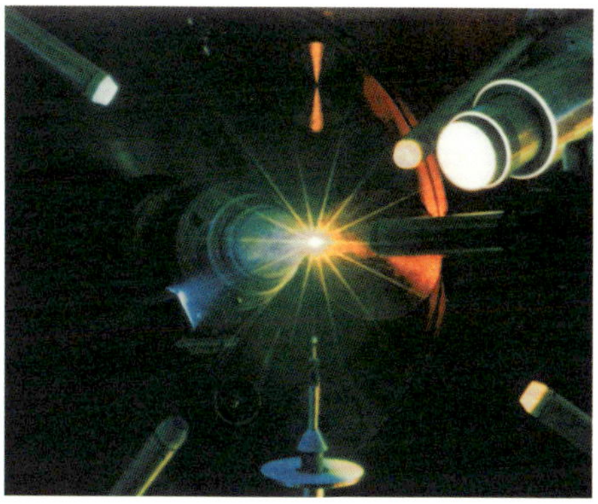

The focal point of the OMEGA laser.

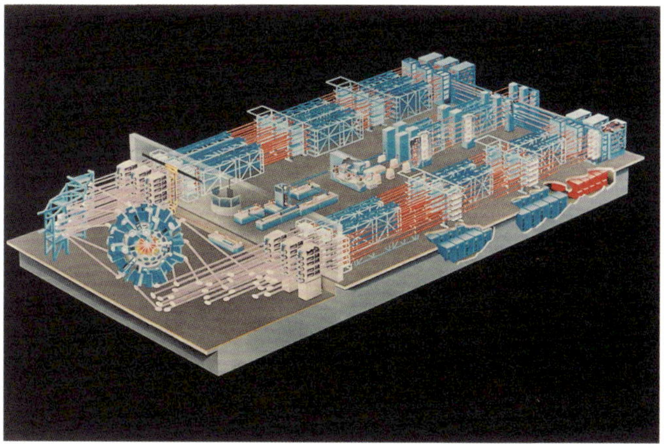

OMEGA laser system.
The enormously powerful OMEGA laser, located at the University of Rochester, New York, is helping scientists learn if an inexhaustible supply of energy can be produced by nuclear fusion. The size of a football field, the OMEGA laser contains 120 optical amplifiers that increase its intensity 100 million times.

boosting the signal

The first long-distance fiber-optic telecommunications systems used electronic amplifiers to strengthen the message-carrying signal. But electronic amplifiers are inefficient. The light must be converted into an electronic signal before it can be amplified. More modern systems—such as those that connect continents under the oceans—use optical amplifiers to boost the signal directly.

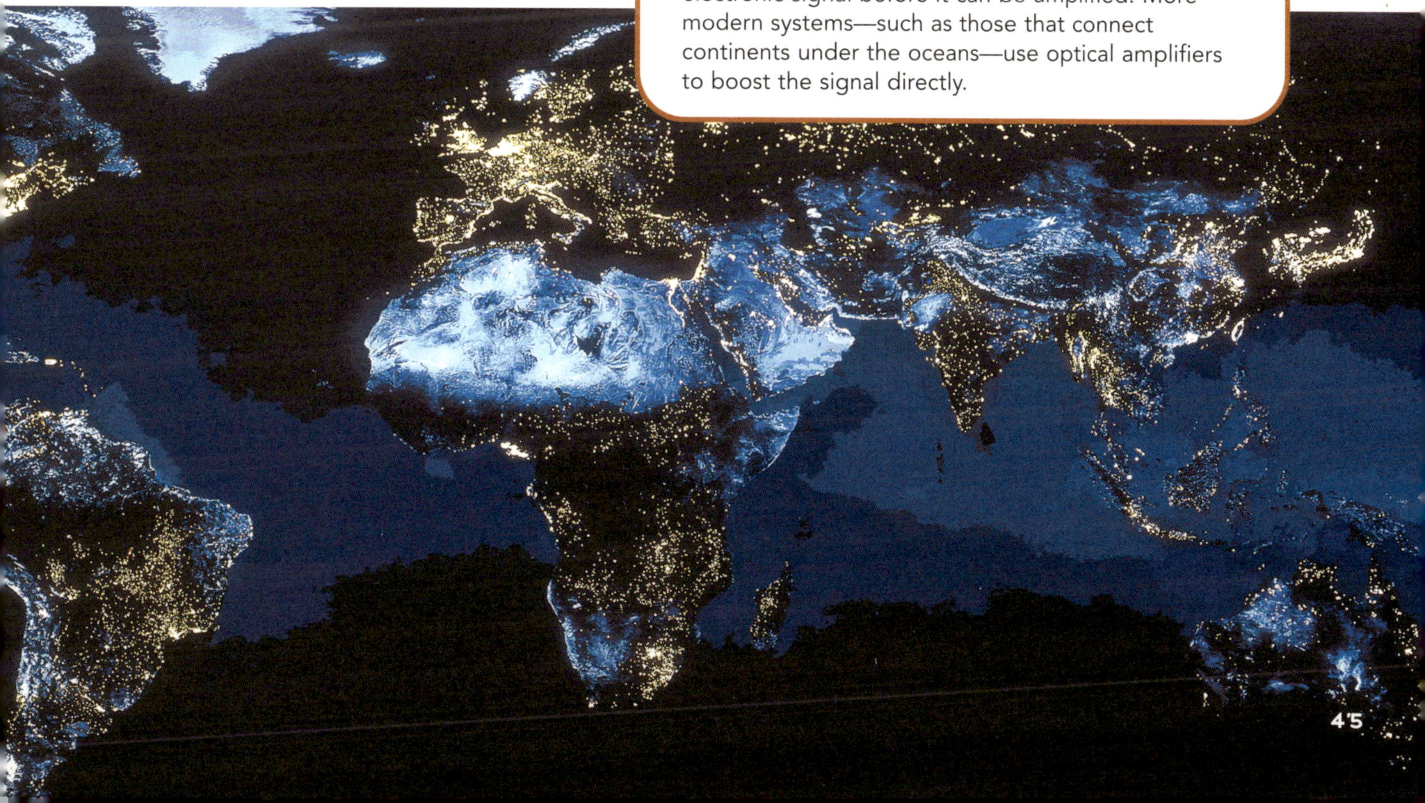

Vessels

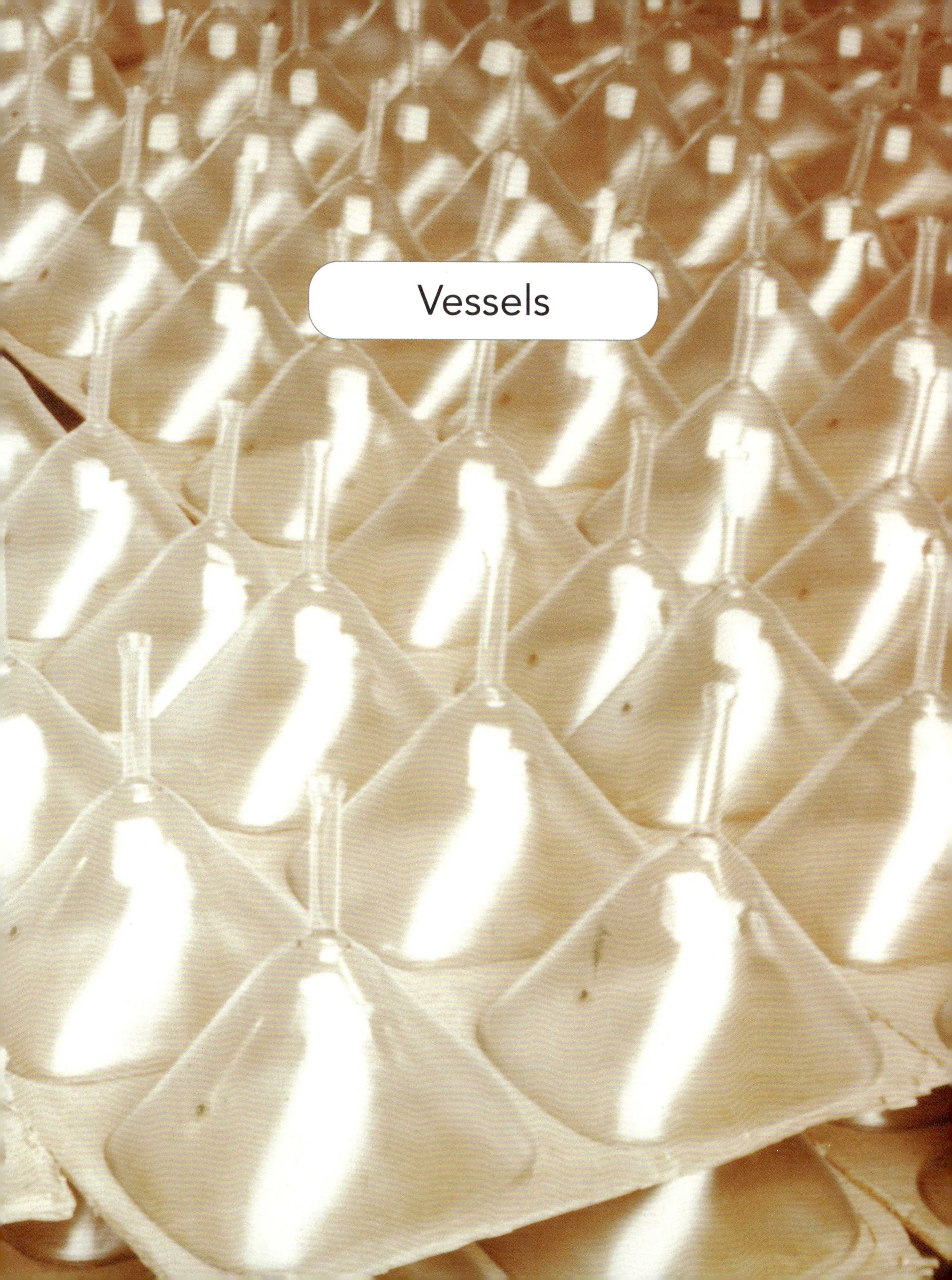

The Iron

By the late 1800s, machines were replacing workers—except in the glass industry. Glassmakers had tried to mechanize glassblowing by imitating the method for making bottles by hand. It didn't work.

PHILIP ARBOGAST 1834–1887

American glassmaker PHILIP ARBOGAST suggested a new approach in 1881. Why not turn the hand method on its head? Glassblowers always finished a bottle's neck and mouth last. If a machine formed them first, it would have something to hold on to while the rest of the bottle was being blown.

Arbogast was right, but he didn't reap the rewards of his revolutionary concept. He sold his patent without building his machine. Others capitalized on his ideas. The machines they built started the glass industry on the road to mechanization. But these semi-automatic machines still needed to have hot glass fed into their molds by hand.

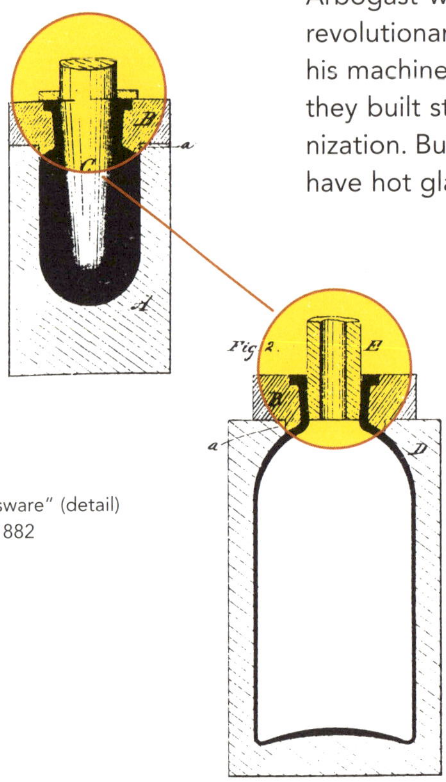

U.S. Patent 260,819
"Manufacture of Glassware" (detail)
P. Arbogast, July 11, 1882

breathing life

A little more than 2,000 years ago, a Roman glassworker did something that dramatically changed the course of history. He blew a puff of air through a hollow rod into a gob of hot glass. The gob inflated into a bubble. At that moment, glassblowing was born.

Glassblowing provided a way to make containers more quickly and cheaply. Glass bottles, once used exclusively to hold rare perfumes and

Life in an early hand shop—ten hours a day, six days a week.

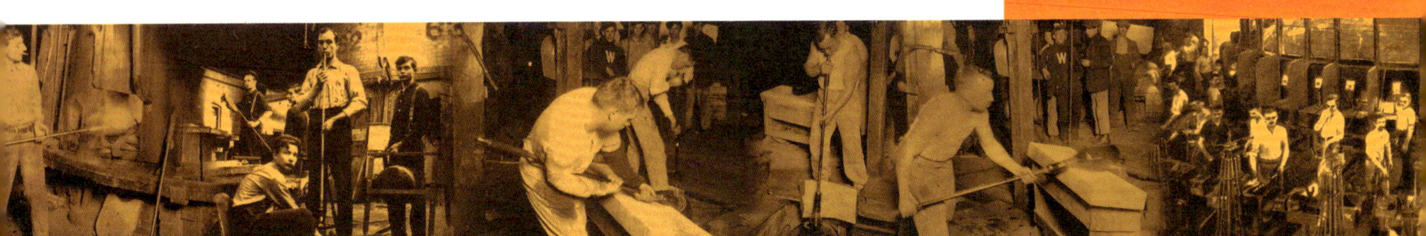

Glassblower

one step closer

almost automatic

Philip Arbogast's invention changed bottle making. He eliminated the glassblower and started the industry on the road to automation. But a person—usually a boy—was still needed to deliver the molten glass to the machine's molds.

Semi-automatic bottle machines based on Arbogast's principle.

into glass

oils, were now available for everyday use. As the demand for bottles soared, shops for making hand-blown glass flourished.

Day after day, in the factories of the mid-1800s, teams of skilled men blew hot glass into molds tended by boys. The work was fast and furious. The factories were hellishly hot. And the demand for glass bottles never stopped.

It was not unusual for children to work in early bottle factories.

the fabulous monster Owens Bottle

MICHAEL OWENS, a self-taught American inventor, propelled the glass industry into the mechanical age. In 1903, he unveiled the world's first completely automatic glass-forming machine—a machine for making bottles.

MICHAEL OWENS
1859–1923

Owens was determined to eliminate the hand delivery of hot glass to bottle machines. He was convinced that a machine could suck thick, molten glass from a rotating pot directly into its molds. This idea defied every glassmaking convention.

It took years of trial and error before Owens introduced his first commercial bottle machine. The mechanical monster had to lift most of its own enormous weight each time it moved a mold to the glass pot. By 1920, the Owens bottle machine had become a 30-ton colossus with 10,000 parts. Even so, it produced more bottles in an hour than a team of human glassblowers could produce in a day.

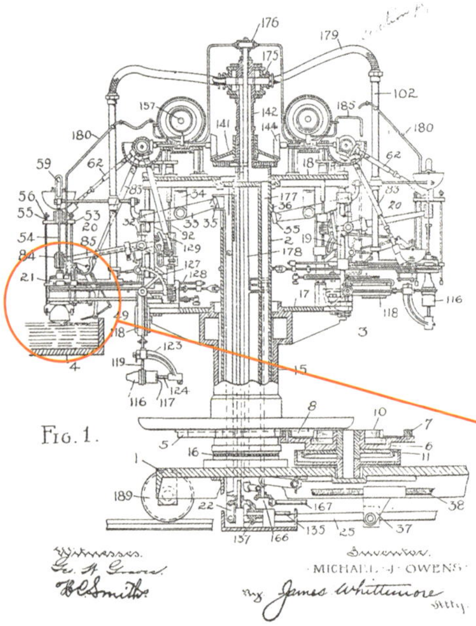

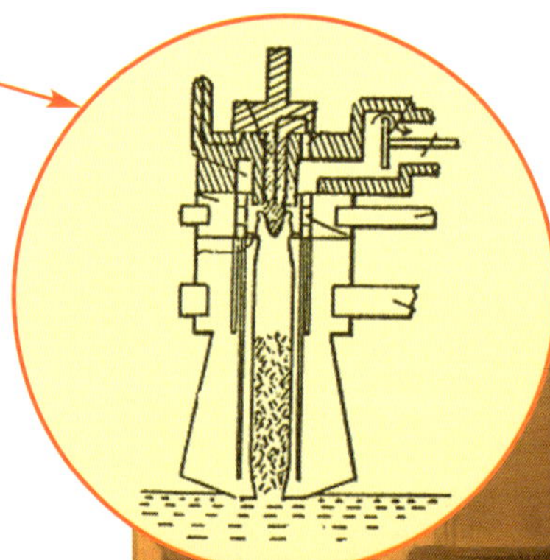

U.S. Patent 766,768
"Glass Shaping Machine" (details)
M. J. Owens, August 2, 1904

Michael Owens *(circled)* and associates standing by Owens's bottle machine, c. 1904.

Machine

The most significant advance in glass production in over 2,000 years ...

American Society of Mechanical Engineers, 1983

The Owens bottle machine.

bottles, bottles, bottles

KARL PEILER
1883–1970

KARL PEILER had no experience with glass, but his invention led to machines that made bottles faster than the Owens machine. The American engineer had been hired by Owens's competitors to find a more practical way to feed glass to bottle machines. In 1915, after months of hazardous experiments with molten glass, Peiler came up with the answer—the gob feeder.

The gob feeder didn't suck up the heavy, molten glass the way the Owens machine did. Instead, it dropped a glass gob of the correct size and shape into the machine's mold. Peiler had learned that the key to forming the gob was controlling the temperature of the glass. If the temperature was right, a paddle could push the needed amount of glass over the rim of the vat at the right time. The falling gob of glass would assume the correct shape naturally.

Peiler's gob feeder was the breakthrough that led to modern glass-forming machines. All the Owens machines are now gone. Today, most of the world's bottles are produced on I.S. (Individual Section) machines, using gob feeders that are successors to Peiler's invention.

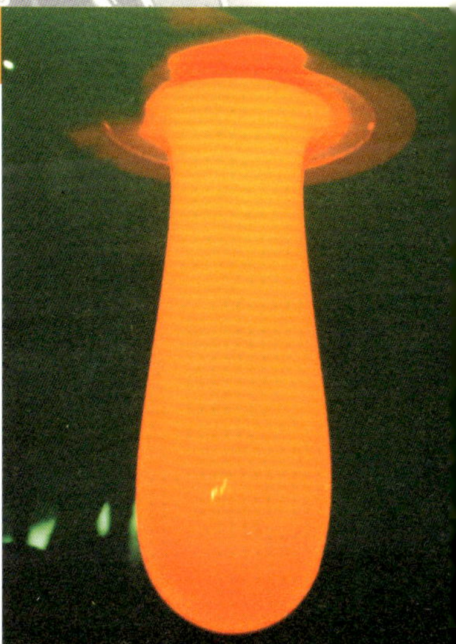

Hot glass gob emerging from a gob feeder

how the I.S. machine works

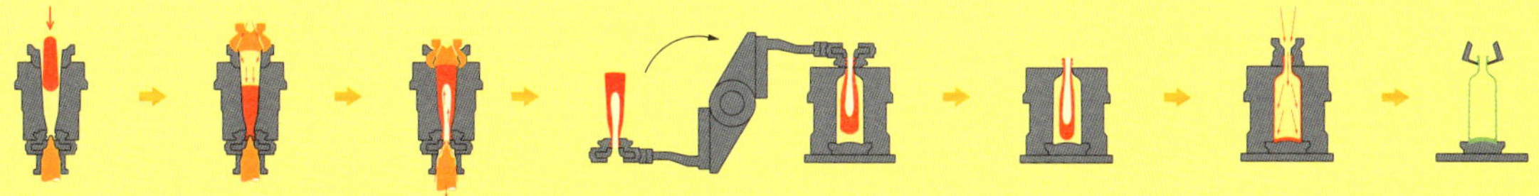

Glass Bulbs & genie in the bottle

In 1879, the brilliant inventor THOMAS EDISON was on the verge of a breakthrough. He had discovered a slow-burning filament that would glow for hours in the vacuum of a glass globe. But to create the first practical electric light, he needed a glass globe that would not implode when the air was drawn out. Edison asked for help.

THOMAS EDISON
1847–1931

WILLIAM WOODS
1879–1937

At Corning Glass Works, glassblowers painstakingly shaped bulbs that could stand up to the new technology. The world's desire for electric light required a more efficient process. A young shop assistant began experimenting, blowing into glass and letting it drop from the end of an iron. Together, air, heat, and gravity shaped the glass into the right bulb form. Glassblowers could now produce thousands of bulbs a day. But they still couldn't keep up. What was needed was a machine to produce thousands of bulbs a minute.

Edison's electric lamp.
Scientific American, 1880.

In 1922, WILLIAM WOODS, a former glassblower, experimented by holding a shovel with a hole under a stream of glass. As the glass sagged through the hole, it formed the perfect shape for blowing a bulb. Woods realized that a ribbon of glass sagged through holes into moving molds could be shaped and blown mechanically into bulbs. The ribbon machine he designed is still state of the art—and still the fastest glass-forming machine ever built.

William Woods's ribbon machine, 1926.

TV Tubes

Light bulbs cascading out of a ribbon machine.

spinning a rectangle

JAMES GIFFEN
1903–1997

The earliest television bulbs were small, round, and made by blowing glass into molds. Later, the bulb's funnel and faceplate were pressed separately, then sealed. As TV-tube bulbs got larger, it became difficult to join the thick, heavy funnels to the faceplates. There was no way to press a thin glass funnel.

At Corning Glass Works, JAMES GIFFEN, a farmer turned glassmaker, had been asked to make a casserole dish by spinning glass in a mold. Giffen soon realized the shape of the casserole was wrong for spinning, but the TV funnel was the right shape. He dropped a glass gob into a spinning mold. Centrifugal force caused the glass to flow up the sides, creating a thin, lightweight funnel.

When the TV industry needed a rectangular bulb, experts said it would be impossible to make the shape by centrifugal spinning. Giffen ignored them. He studied how the glass flowed, redesigned the mold, and spun the glass into the new shape.

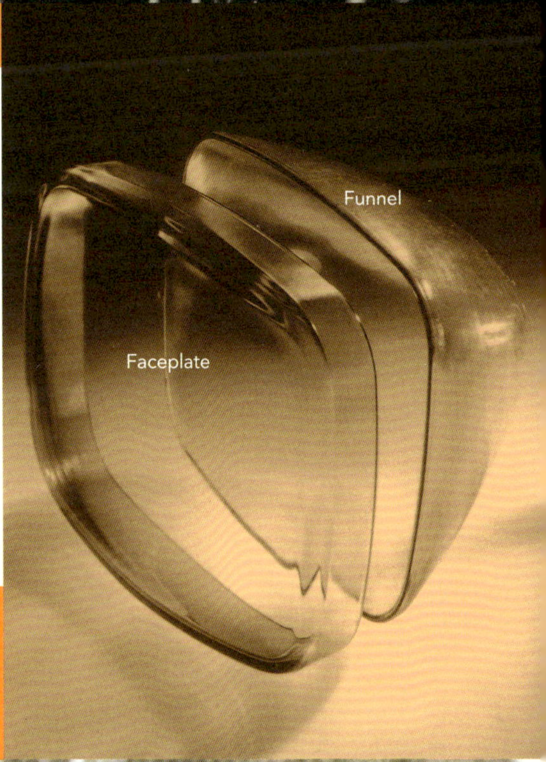

The faceplate and funnel of a TV bulb are made separately, then sealed. In 1949, James Giffen found a way to form a funnel by spinning glass in a mold. Centrifugal spinning is no longer used to make thin glass funnels because pressing technology has become sophisticated enough to produce them.

Borosilicate Glass

finding the right recipe

OTTO SCHOTT, the pioneering German glass chemist, made a glass that could reliably do something that didn't seem possible: endure sudden, uneven temperature shifts without shattering. The key, Schott discovered in 1882, was to have a critical amount of the element boron in the glass recipe. Schott's new durable, heat-resistant borosilicate glass was an immediate commercial success.

OTTO SCHOTT 1851–1935

W. C. TAYLOR 1886–1958

EUGENE SULLIVAN 1872–1962

Building on Schott's work, Corning chemist W. C. TAYLOR found the right recipe to make an even more temperature-tolerant borosilicate glass. With it, railway signal lanterns didn't shatter in sleet or freezing rain. By 1915, Taylor and chemist EUGENE SULLIVAN had transformed borosilicate glass into an exotic new product—Pyrex® glass cookware. It captured America's imagination. Soon, in kitchens all over the country, people were cooking in glass.

Schott's and Corning's work launched the era of modern glass chemistry research. Today, there is a "cookbook" of more than a million different glass recipes, each precisely formulated to meet a product need.

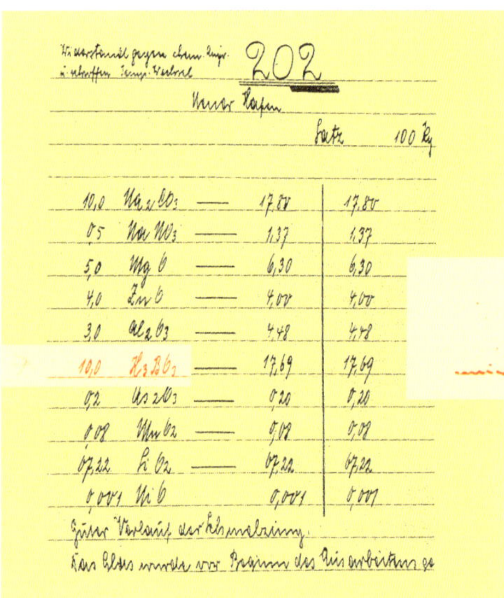

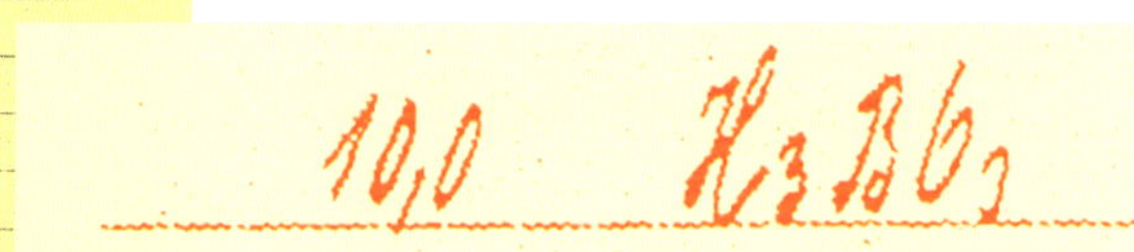

Detail from Otto Schott's batch notebook showing a formula for borosilicate glass, c. 1891.

Low-expansion borosilicate glass was developed in Germany to make thermometers that would resist changing their volume with temperature changes. In America, the new glass was first used for railroad signal lanterns that wouldn't shatter with changes in the weather.

Pyrex® flask.
Labware is still made by hand out of borosilicate glass. It resists chemical corrosion and temperature changes that take place during experiments.

good vibrations

Heat causes atoms in glass to vibrate—they alternately move farther apart, then closer together. The higher the heat, the greater the vibrations. If the vibrating atoms move away from each other more than they move together, the glass expands. When glass is heated unevenly, some parts expand more than others, and the glass usually shatters. The element boron can prevent that.

In the right amount, boron moderates the vibrations. It makes the distance between atoms grow and shrink by almost the same amount; the net movement of the atoms is nearly zero. There's little expansion, so the glass doesn't shatter.

baking a new business

When physicist JESSE LITTLETON joined Corning Glass Works in 1914, the company was using its new heat-resistant glass to make battery jars and signal lanterns. Littleton's job was to evaluate the properties of the new glass. One day, his wife's ceramic casserole dish cracked. Bessie suggested that the heat-resistant glass might be just right for baking. Jesse went to the lab and sawed off a battery jar. He gave it to his wife, who successfully baked a cake in it.

Bessie continued baking in battery jars, and even made custards in lamp chimneys. Her kitchen experiments provided the impetus for a new consumer-product business—see-through Pyrex® glass cookware.

Left: Jesse and Bessie Littleton, c. 1912.
Right: Advertisement for Pyrex® glass dishes, 1916

Glass-Ceramics
a discovery waiting to happen

Crystals are usually a glassmaker's enemy. When they form in glass, crystals can change the properties of the material in unwanted ways. Corning chemist DONALD STOOKEY saw crystals differently.

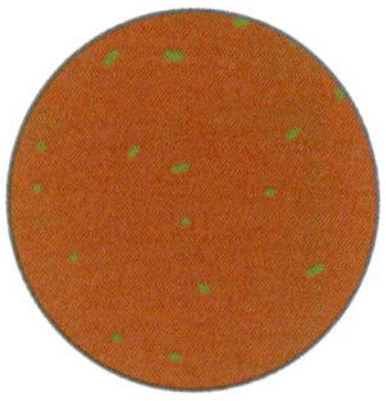

1 Nuclei Formation

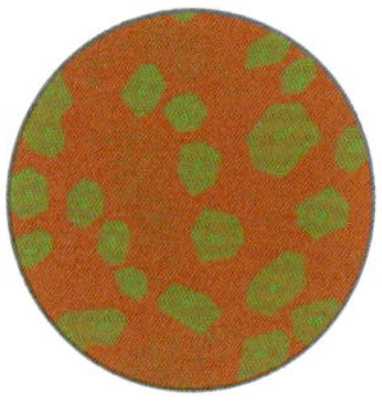

2 Crystal Growth on Nuclei

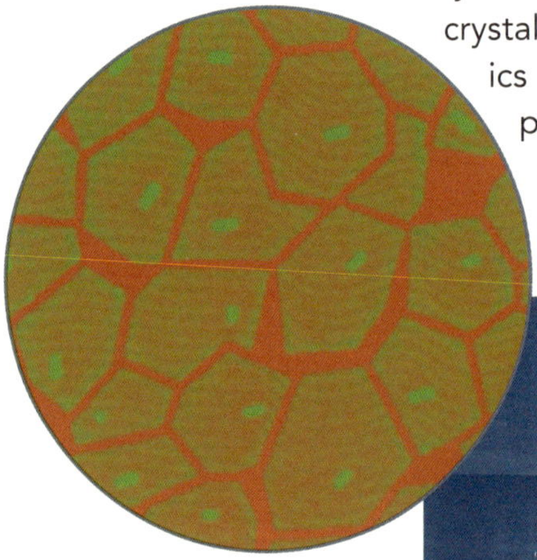

3 Glass-Ceramic Microstructure

DONALD STOOKEY b. 1915

One night in 1954, he put a piece of experimental glass into a furnace. The next morning, he discovered that the furnace had overheated. He was sure he would find a pool of melted glass inside. Instead, he found an opaque solid that was so strong it bounced when he accidentally dropped it. Microscopic crystals had formed, changing the glass into a new material that Stookey called a glass-ceramic.

Stookey knew he could use heat and chemistry to control crystallization. Maybe it would be possible to form different crystals in different glasses. That would mean glass-ceramics could be tailored to have a remarkable range of properties. Stookey was right. Today, glass-ceramics are used to make everything from heat-defying cookware to exotic optical devices.

a recipe for glass-ceramic

1. Make the glass, adding a special material—a nucleating agent—to the raw ingredients of the glass batch. The nucleating agent forms microscopic particles called nuclei.
2. Heat the glass again. Crystals will grow on the nuclei.
3. The result is a new material, a glass-ceramic, with properties different from the original glass.

Missile nose cones, or radomes, start out as "green" glass. They are reheated to form a glass-ceramic that is strong, heat-resistant, and, most important, transparent to the radar that guides the missile.

Why did such an important discovery occur so late in the ... history of glass, and why was an accident necessary to bring it about?

Donald Stookey, 1977

Cylindrical shell for AXAF orbiting X-ray telescope mirror, Zerodur® glass-ceramic. The six concentric cylindrical shells that make up the mirror for the AXAF telescope were designed to collect as much X-ray radiation as possible. The shells are made from a glass-ceramic that resists distortion with changes in temperature.

from glass to glass-ceramic

The secret to making a glass-ceramic is control. Scientists select the exact temperatures and ingredients needed to make crystals grow in the glass. When enough crystals have formed, the glass becomes a new material: glass-ceramic. These images *(top to bottom)* show mica crystals forming in a glass as the temperature increases.

Fiberglass

birth of a new industry

The world's largest bottle plant stood empty. Prohibition and the Depression had crushed the demand for bottles. American bottle maker Owens-Illinois needed a new product. There might be an enormous market for insulation made of lightweight glass fibers, but no one had found a way to mass-produce fibers that were fine enough.

DALE KLEIST
1908–1998

Owens-Illinois tried drawing fibers by using air and steam, but failed to produce anything thinner than broom straws. Then, in 1932, a lucky accident happened. DALE KLEIST, an intern engineer, was using a metal-layer gun—sometimes used to bronze baby shoes—to try to generate a stream of glass to fuse two halves of a glass block. Instead of a stream, the jets of air coming from the gun produced an explosion of fine, lightweight fibers.

Kleist's frustration was Owens-Illinois's salvation. His "experiment" showed that air or steam, applied at a critical angle, can stretch glass into very fine fibers. Within months, Owens-Illinois had converted the disused bottle plant into a factory for manufacturing fiberglass wool insulation.

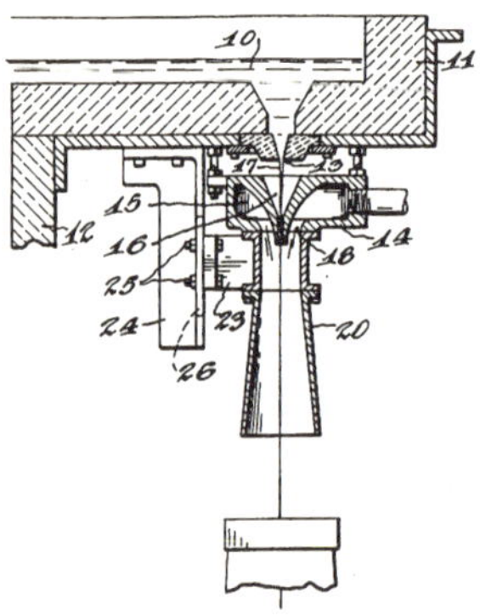

U.S. Patent 2,121,802
"Method and Apparatus for Strengthening Fibers" (detail)
D. Kleist and John H. Thomas, June 28, 1938

like cotton candy

It looks a lot like making cotton candy—but the modern rotary method for making glass wool insulation took much longer to perfect. Like the confectioner's process, it uses centrifugal force to rapidly transform a viscous liquid into fluffy fibers.

German inventors FRITZ HAGER and FRIEDRICH ROSENGARTH patented a method in 1929 that used centrifugal force to mass-produce glass fiber. A stream of molten glass fell onto a rotating disk and was spun off as droplets that trailed long filaments. The filaments were fairly fine, but they clung to the disk's drive shaft and had to be sawed off.

In the 1950s, Owens Corning and France's St.-Gobain refined and combined the rotary method and the blowing process. As glass passed through holes in the rim of a rotating disk, a blast of hot air further stretched the filaments, making them even finer and more lightweight. This fast, efficient method continues to meet an ever-growing demand. Fiberglass insulation was once rare. Today, it's rare to find a new American home without it.

Installing fiberglass insulation in a ship, c. 1940.

> If necessity is the mother of invention, then for the glass fiber industry, adversity is the father.
>
> William Boeschenstein, Owens-Corning Fiberglas, 1995

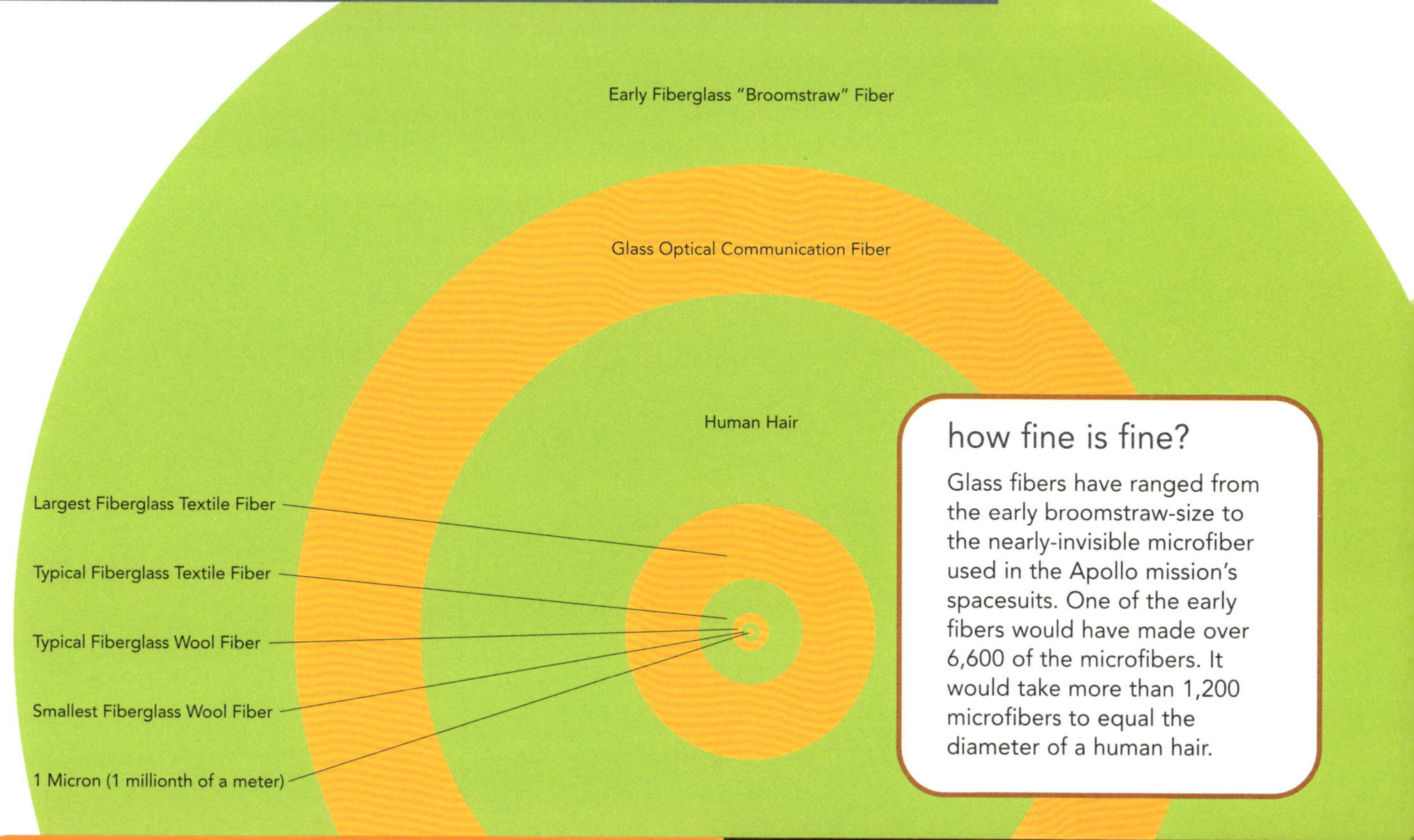

how fine is fine?

Glass fibers have ranged from the early broomstraw-size to the nearly-invisible microfiber used in the Apollo mission's spacesuits. One of the early fibers would have made over 6,600 of the microfibers. It would take more than 1,200 microfibers to equal the diameter of a human hair.

flexible as silk

A glass dress? The mere thought drew crowds of people to Chicago's Columbian Exposition in 1893. They had hoped the dress would be transparent. It wasn't. And it was barely wearable. Its glass fibers broke when they were bent or rubbed, so the dress was prickly. If glass fibers were to be as flexible as silk, they would have to be a lot finer.

German innovator OSCAR GOSSLER tried drawing finer fibers in 1908, pulling them over a drum from streams of melted glass that emerged from holes in a vat. The trouble was that the corrosive glass eventually enlarged the holes, and the fibers became coarse.

By lining the holes with corrosion-resistant platinum and using high-quality glass, Owens Corning produced fiberglass filaments so fine that they were flexible enough to be spun into thread. Today, the thread is used not for clothing but for making industrial fabrics and adding strength and durability to many plastics.

The fiberglass spacesuits worn by the first men on the moon were a giant leap from the dress on display at the Chicago world's fair in 1893.

Index

pp. 8–9

float tank interior: Collection of The Corning Museum of Glass

p. 10

crown glass being spun: source unknown

crown glass cutting illustration: Courtesy Pilkington plc

p. 11

Crystal Palace: Collection of the Juliette K. and Leonard S. Rakow Research Library of The Corning Museum of Glass

cylinder windowpane illustration: Ralph Appelbaum Associates

p. 12

Emile Fourcault: Collection of the Juliette K. and Leonard S. Rakow Research Library of The Corning Museum of Glass

updraw machine diagram: Ralph Appelbaum Associates

twin-grinding machinery: Courtesy Pilkington plc

p. 13

cylinder factory interior: Courtesy Pilkington plc

p. 14

Sir Alastair Pilkington: Courtesy Pilkington plc

float process diagram: Ralph Appelbaum Associates

p. 15

float tank interior: Collection of The Corning Museum of Glass

glass emerging from annealer: Collection of The Corning Museum of Glass

p. 16

Prince Rupert: Photo AKG London

p. 17

Gemini space capsule: Courtesy NASA

chem-tempering diagram: Courtesy Corning Incorporated, Corporate Archives

Jesse Littleton: Courtesy the Littleton Family

Prince Rupert drop illustration: Ralph Appelbaum Associates

exploding drop: Photo by James L. Amos

p. 18

Edouard Bénédictus: Courtesy Archives de Saint-Gobain

TripleX™ advertisement: Courtesy Archives de Saint-Gobain

lamination illustration: Ralph Appelbaum Associates

p. 19

early auto accident: Courtesy Archives de Saint-Gobain

crash-test series: Courtesy GM Corporation, ©1994, used with permission of GM Media Archives

p. 20

1934 Chrysler Airflow: Reprinted with permission of the Chrysler Historical Collection

p. 21

press-bending assembly line: Courtesy Pilkington plc

gravity-shaping illustration: Ralph Appelbaum Associates

press-bending illustration: Ralph Appelbaum Associates

EV1 electric car: Courtesy GM Corporation, Advanced Technology Vehicles

p. 22

Stuart Dockerty: Courtesy Corning Incorporated, Corporate Archives

Clint Shay: Courtesy Corning Incorporated, Corporate Archives

fusion draw illustration: Ralph Appelbaum Associates

p. 23

float/fusion surfaces: Courtesy Corning Incorporated, Advanced Display Products

laptop computer: Courtesy 1999 PhotoDisc, Inc.

sunglasses: Courtesy Corning Incorporated

Javelin car: Reprinted with permission of the Chrysler Historical Collection

LCD matrix: Courtesy Corning Incorporated, Advanced Display Products

p. 24

Roy Gordon: Courtesy Roy Gordon

CVD illustration: Ralph Appelbaum Associates

p. 25

blazing sun: Courtesy Lockheed Martin Corporation

Bank of China: Corbis/Bettmann

pp. 26–27

optical-fiber manufacturing: Courtesy Corning Incorporated

p. 28

Hans Lippershey: Corbis/Bettmann

Antonj van Leeuwenhoek: Corbis/Bettmann

Leeuwenhoek microscope: Corbis/Bettmann

Lippershey telescope diagram: Ralph Appelbaum Associates

truffle spores: Brian J. Ford, 1997

p. 29

Galileo illustration: Image Select/Art Resource, NY

Galileo telescope: Corbis/Bettmann

light-bending diagram: Ralph Appelbaum Associates

complex-microscope diagram: Ralph Appelbaum Associates

Hooke microscope: Corbis/Bettmann

p. 30

Sir Isaac Newton: Corbis/Bettmann

Newton telescope: e.t. archive

Porter rendering: Courtesy the California Institute of Technology, Institute Archives

George Hale: Courtesy the Yerkes Observatory

Photo Credits

p. 31

refracting telescope diagram: Ralph Appelbaum Associates

reflecting telescope diagram: Ralph Appelbaum Associates

200" telescope mirror blank: Courtesy Corning Incorporated, Corporate Archives

p. 32

Otto Schott: Courtesy Schott Glas, Mainz, Germany

Ernst Abbe: Courtesy Schott Glas, Mainz, Germany

Newton with prism: Corbis/Bettmann

chromatic aberration grid: Dorling Kindersly Ltd., Courtesy the Science Museum, London

p. 33

periscope cutaway: Courtesy Kollmorgen Electro-Optical Corporation

lens-stacking diagram: Ralph Appelbaum Associates

p. 34

Charles DeVoe: Courtesy Corning Incorporated, Corporate Archives

flawed glass: Courtesy Corning Incorporated, Corporate Archives

p. 35

optical-melting process diagram: Ralph Appelbaum Associates

German glassworker: Courtesy Pilkington plc

p. 36

J. Franklin Hyde: Courtesy J. Franklin Hyde

fused-silica process diagram: Ralph Appelbaum Associates

space shuttle: Courtesy Corning Incorporated, Corporate Archives

stepper lens illustration: Ralph Appelbaum Associates

p. 37

Hubble telescope mirror: Courtesy Raytheon Optical Systems Inc.

p. 38

Augustin Fresnel: Corbis/Bettmann

Makapuu Lighthouse lens: Courtesy U.S. Coast Guard Historian's Office

Fresnel lens illustration: Ralph Appelbaum Associates

p. 39

Amelia Island lighthouse: Courtesy Ed Matthews, 1988

railroad lantern: Courtesy Corning Incorporated

p. 40

Heinrich Lamm: Courtesy Michael Lamm

Lamm experiment: Courtesy Michael Lamm

fiber-bundle illustration: Ralph Appelbaum Associates

Lawrence Curtiss: Courtesy Lawrence Curtiss

coating illustration: Ralph Appelbaum Associates

p. 41

J. Daniel Colladon: Special Collections Department, Brandeis University Libraries

John Tyndall: Corbis/Bettmann

Colladon's fountain: Trustees of the Boston Public Library

total internal reflection illustration: Ralph Appelbaum Associates

p. 42

Donald Keck: Courtesy Corning Incorporated, Corporate Archives

Robert Maurer: Courtesy Corning Incorporated, Corporate Archives

Peter Schultz: Courtesy Corning Incorporated, Corporate Archives

Keck's notebook: Courtesy Corning Incorporated, Technical Information Center

p. 43

optical fiber being drawn: Courtesy Corning Incorporated

outside vapor deposition: Courtesy Corning Incorporated

Bell's photophone: Corbis/Bettmann

p. 44

Elias Snitzer: Courtesy Elias Snitzer

optical amplifier: Courtesy Corning Incorporated

optical amplifier diagram: Ralph Appelbaum Associates

p. 45

telecommunication map: Courtesy Corning Incorporated

OMEGA laser system: Courtesy Laser Energetics Laboratory, University of Rochester

laser focal point: Courtesy Laser Energetics Laboratory, University of Rochester

pp. 46–47

television tubes: Courtesy Corning Incorporated, Corporate Archives

p. 48

Philip Arbogast: Courtesy Maxine Arbogast and Ken Hodges

hand-shop photo collage: Courtesy Corning Incorporated, Corporate Archives, and the Collection of the Juliette K. and Leonard S. Rakow Research Library of The Corning Museum of Glass

p. 49

glassblowers: Courtesy Corning Incorporated, Corporate Archives

p. 50

Michael Owens: Courtesy Owens-Illinois, Inc.

men standing by Owens machine: Courtesy Owens-Illinois, Inc.

p. 51

Owens bottle machine: Courtesy Owens-Illinois, Inc.

Karl Peiler: Courtesy Emhart Powers

glass gob: Courtesy Corning Incorporated, Corporate Archives

I.S. machine diagram: Ralph Appelbaum Associates

p. 52

Thomas Edison: Courtesy the U.S. Department of the Interior, National Park Service, Edison National Historic Site

William Woods: Courtesy Corning Incorporated, Corporate Archives

Edison lamp illustration: Courtesy Smithsonian Institution, Neg. # 47961B

ribbon machine: Courtesy Corning Incorporated, Corporate Archives

p. 53

light bulbs emerging from machine: Courtesy Corning Incorporated, Corporate Archives

James Giffen: Courtesy Corning Incorporated, Corporate Archives

television-bulb parts: Courtesy Corning Incorporated, Corporate Archives

p. 54

Otto Schott: Courtesy Schott Glas, Mainz, Germany

W. C. Taylor: Courtesy Corning Incorporated, Corporate Archives

Eugene Sullivan: Courtesy Corning Incorporated, Corporate Archives

Schott's notebook: Courtesy Schott Glas, Mainz, Germany

railroad lantern: Courtesy The Rockwell Collection

Jesse and Bessie Littleton: Courtesy the Littleton Family

p. 55

Pyrex® flask and flame: Collection of The Corning Museum of Glass

Pyrex® advertisement: Courtesy Corning Incorporated, Corporate Archives

p. 56

Donald Stookey: Courtesy Corning Incorporated, Corporate Archives

glass-ceramic illustration: Courtesy George Beall

man with radomes: Courtesy Corning Incorporated, Corporate Archives

p. 57

AXAF X-ray telescope: Courtesy Raytheon, Danbury, Connecticut

glass-ceramic micrographs: Courtesy George Beall

p. 58

Dale Kleist: Courtesy Mr. and Mrs. Dale Kleist

installing fiberglass insulation: Courtesy Owens Corning

p. 59

relative fiberglass sizes illustration: Ralph Appelbaum Associates

fiberglass dress: Courtesy Owens-Illinois, Inc.

astronaut: Courtesy NASA